Particulates and Continuum-Multiphase Fluid Dynamics

Particulates and Continuum-Multiphase Fluid Dynamics

Contributors

Rabindra Nath Mondal, Shinichiro Yanase et al.

AURIS Reference

www.aurisreference.com

Particulates and Continuum-Multiphase Fluid Dynamics

Contributors: Rabindra Nath Mondal, Shinichiro Yanase et al.

Published by Auris Reference Limited

www.aurisreference.com

United Kingdom

Particulates and Continuum-Multiphase Fluid Dynamics

ISBN: 978-1-78154-538-6

British Library Cataloguing in Publication Data
A CIP record for this book is available from the British Library

Printed in the United Kingdom

Contents

List of Contributors

Rabindra Nath Mondal
Department of Mathematics, Jagannath University, Dhaka, Bangladesh

Samir Chandra Ray
Department of Mathematics, Bangabandhu Sheikh Mujibur Rahman Science and Technology University, Gopalganj, Bangladesh

Shinichiro Yanase
Department of Mechanical and System Engineering, Faculty of Engineering, Okayama University, Okayama, Japan

Li Li
Aeronautics Computing Technique Research Institute (ACTRI), P.O. Box 90, Xi'an 710068, China

W. Limtrakarn
Department of Mechanical Engineering, Thammasat University, Pathumthani 12120, Thailand

P. Boonmongkol
Department of Mechanical Engineering, Thammasat University, Pathumthani 12120, Thailand

A. Chompupoung
Royal Project Foundation, Chiang Mai 50200, Thailand

K. Rungprateepthaworn
National Electronics and Computer Technology Center, Klong Luang, Pathumthani 12120, Thailand

J. Kruenate
National Metal and Materials Technology Center, Klong Luang, Pathumthani 12120, Thailand

P. Dechaumphai5
Department of Mechanical Engineering, Chulalongkorn University, Bangkok 10330, Thailand

Vadim Seleznev
Physical and Technical CenterSarov, Nizhny Novgorod Region,, Russia

Carlos Gavilán Moreno
Cofrentes N.P.P. Iberdrola S.A.,Spain

Ki-Pyo You
Department of Architecture Engineering, Chonbuk National University, 567 Baekje-daero, deokjin-gu, Jeonju-si, Jeollabuk-do 561-756, Republic of Korea

Young-Moon Kim
Department of Architecture Engineering, Chonbuk National University, 567 Baekje-daero, deokjin-gu, Jeonju-si, Jeollabuk-do 561-756, Republic of Korea

Mohammad Bakhtiari
Department of Ocean Engineering, Amirkabir University of Technology, Tehran, Iran

Hassan Ghassemi
Department of Ocean Engineering, Amirkabir University of Technology, Tehran, Iran

Alessandro Cubeddu
Institute of Fluid Mechanics, Technical Faculty, University of Erlangen-Nuremberg, Erlangen, Germany
Department of Food Biotechnology and Food Process Engineering, Faculty III—Process Science and Engineering, Berlin Institute of Technology, Berlin, Germany

Cornelia Rauh
Institute of Fluid Mechanics, Technical Faculty, University of Erlangen-Nuremberg, Erlangen, Germany
Department of Food Biotechnology and Food Process Engineering, Faculty III—Process Science and Engineering, Berlin Institute of Technology, Berlin, Germany

Antonio Delgado
Institute of Fluid Mechanics, Technical Faculty, University of Erlangen-Nuremberg, Erlangen, Germany

Sooraj K. Prabha
Computational Nanotechnology Laboratory, National Institute of Technology Calicut, Kozhikode-673 601, Kerala, India

Sreehari P. D.
Computational Nanotechnology Laboratory, National Institute of Technology Calicut, Kozhikode-673 601, Kerala, India

Murali Gopal M.
Computational Nanotechnology Laboratory, National Institute of Technology Calicut, Kozhikode-673 601, Kerala, India

Sarith P. Sathian
Computational Nanotechnology Laboratory, National Institute of Technology Calicut, Kozhikode-673 601, Kerala, India

Claus Kiefer
Support Center for Advanced Neuroimaging, Institute of Diagnostic and Interventional Neuroradiology, University of Bern (Inselspital), Bern, Switzerland

Frauke Kellner-Weldon
Support Center for Advanced Neuroimaging, Institute of Diagnostic and Interventional Neuroradiology, University of Bern (Inselspital), Bern, Switzerland

Preface

Treating multiphase systems with emphasis on the aspect of fluid dynamics and as an introduction to research in multiphase flow, covers definitive concepts, methods, and theories which have been validated by experimental results. Combined effects of centrifugal and coriolis instability of the flow through a rotating curved duct with rectangular cross section have been discussed in first chapter. Second chapter focuses on initial investigation of adjoint-based unstructured grid adaptation for vortical flow simulations. Third chapter deals with computational fluid dynamics modeling to improve natural flow rate and sweet pepper productivity in greenhouse. Fourth chapter explores the application of computational fluid dynamics to practical design and performance analysis of turbomachinery. Turbulence, vibrations, noise and fluid instabilities have been presented in fifth chapter. Sixth chapter determines the internal wind flow inside a windbreak fence, identifying an area unaffected by wind in order to measure the snowfall. The aim of seventh chapter is to discuss the numerical analysis of electromagnetic control of the boundary layer flow on a ship hull. Eighth chapter gives the information of 3D thermo-fluid dynamic simulations of high-speed-extruded starch based products. The purpose of ninth chapter is to explain the effect of system boundaries on the mean free path for confined gases. Last chapter describes the computational fluid dynamics and its impact on flow measurements.

Chapter 1

COMBINED EFFECTS OF CENTRIFUGAL AND CORIOLIS INSTABILITY OF THE FLOW THROUGH A ROTATING CURVED DUCT WITH RECTANGULAR CROSS SECTION

Rabindra Nath Mondal[1], Samir Chandra Ray[2], Shinichiro Yanase[3]

[1]Department of Mathematics, Jagannath University, Dhaka, Bangladesh

[2]Department of Mathematics, Bangabandhu Sheikh Mujibur Rahman Science and Technology University, Gopalganj, Bangladesh

[3]Department of Mechanical and System Engineering, Faculty of Engineering, Okayama University, Okayama, Japan

ABSTRACT

Combined effects of centrifugal and coriolis instability of the flow through a rotating curved duct with rectangular cross section have been studied numerically by using a spectral method, and covering a wide range of the Taylor number $0 \leq Tr \leq 2500$ for a constant Dean number $Dn = 2000$. The rotation of the duct about the center of curvature is imposed in the positive direction, and the effects of rotation (Coriolis force) on the flow characteristics are investigated. As a result, multiple branches of asymmetric steady solutions with two-, three- and multi-vortex solutions are obtained. To investigate the non-linear behavior of the unsteady solutions, time evolution calculations as well as power spectrum of the unsteady solutions are performed, and it is found that the unsteady flow undergoes through various flow instabilities in the scenario "chaotic → multi-periodic → periodic → steady-state", if Tr is increased in the positive direction. The present results show the characteristics of both the secondary flow and axial flow distribution in the flow.

INTRODUCTION

Recently, great attention has been paid for the study of flows and heat transfer through rotating curved ducts and channels because of its practical application in chemical, mechanical, bio-mechanical and biological engineering. A quantitative analogy between flows in stationary curved pipes and orthogonally rotating straight pipes has been reported by Ishigaki [1] [2]. Taking this analogy as a basis, this study describes the characteristics of more general and complicated flow in rotating curved ducts. Such rotating flow passages are used in cooling systems in rotating machinery such as in gas turbines, turbo-machinery, electric generators and electric motors. The readers can refer to Berger et al. [3] and Nandakumar and Masliyah [4] for some outstanding reviews on curved duct flows.

One of the interesting phenomena of the flow through a curved duct is the bifurcation of the flow because generally there exist many steady solutions due to channel curvature. An early complete bifurcation study of two-dimensional (2-D) flow through a curved duct with square cross section was performed by Winters [5]. However, an extensive treatment of the flow through a curved square duct was reported by Mondal et al. [6]. He found a close relationship between the unsteady solutions and the bifurcation diagram of steady solutions. Ishigaki [2] examined the flow structure and friction factor numerically for both the counter-rotating and co-rotating curved circular pipe with a small curvature. Selmi et al. [7] examined the combined effects of system rotation and curvature on the bifurcation structure of two-dimensional flows in a rotating curved square duct. Selmi and Nandakumer [8] performed studies on the flow in rotating curved rectangular ducts. Yamamoto et al. [9] employed spectral method to examine the flow structure and the flow rate ratio in a rotating curved square duct flow and found six-cell phenomenon of the secondary flow. Recently, Mondal et al.

[10] performed a comprehensive numerical study on the bifurcation structure and the stability of solutions for laminar forced convection in a rotating curved duct of square cross section, and revealed some new features on fluid flow. However, the complete flow structures under the combined action of coriolis and centrifugal instability are still unrevealed for the rotating curved rectangular duct flow at high rotation. It is an attempt of the present study.

It is well known that, fluid flowing in a rotating curved duct is subjected to two forces: the Coriolis force, caused by the rotation of the duct, and centrifugal force caused by the curvature of the duct. These two forces affect each other, as a result complex behavior of the secondary flow and the axial flow can be obtained (Wang and Cheng [11]). For isothermal flows of a constant property fluid, however, the Coriolis force tends to produce vorticity while centrifugal force is purely hydrostatic (Zhang et al. [12]). When a temperature induced variation of fluid density occurs for non-isothermal flows, both Coriolis and centrifugal type buoyancy forces can contribute to the generation of vortices (Mondal et. al., [13]). These two effects of rotation either enhance or counteract each other in a non-linear manner depending on the direction of wall heat flux and the flow domain. Therefore, the effect of rotation of the system is more subtle and complicated and yields new; richer features of flow and heat transfer in general, bifurcation and stability in particular, for non-isothermal flows. Mondal et al. [13]] performed numerical prediction of the non-isothermal flows through a rotating curved square duct and revealed some of such new features. Very recently, Mondal et al. [14] investigated the combined effects of the Coriolis force and the centrifugal force on the flows in a rotating curved square duct numerically. The secondary flow characteristics in a curved square duct were investigated experimentally by using visualization method by Yamamoto et al. [15]. Three-dimensional incompressible viscous flow and heat transfer in a rotating U-shaped square duct were studied numerically by Nobari et al. [16]. However, there is no known study on bifurcation and unsteady flow characteristics in a rotating curved rectangular duct with large rotational speed. The present paper is, therefore, an attempt to fill up this gap.

Time dependent analysis of fully developed curved duct flows was first initiated by Yanase and Nishiyama [17] for a rectangular cross section. In that study, they investigated unsteady solutions for the case where dual solutions exist. However, time-dependent behavior of the flow in a curved rectangular duct over a wide range of aspect ratios was investigated, in detail, by Yanase et al. [18] numerically. They observed that periodic oscillations are available with symmetry condition while aperiodic time evolutions without symmetric condition. Wang and Yang [19] [20] performed numerical as well as experimental investigation on fully developed periodic oscillation in a curved square duct. Flow visualization in the range of Dean Numbers from 50 to 500 was carried out in their experiment. They showed, both experimentally and numerically, that the temporal oscillation takes place between symmetric asymmetric 2-cell and 4-cell flows where there are no stable steady flows. Applying spectral method, Yanase et al. [21] performed comprehensive numerical study of the time-dependent solutions for the non-isothermal flows through a curved rectangular duct, and studied the effects of secondary flows on convective heat transfer. In order to study the time-dependent behavior of the unsteady solutions, recently, Mondal et al. [22] performed numerical prediction of the unsteady solutions through curved square duct for isothermal flow. They showed that periodic solutions turn into chaotic solution through a multi-periodic solution, if the Dean number is increased no matter what the curvature is. However, transient behavior of the unsteady solutions, such as periodic, multi-periodic or chaotic solutions, is yet unresolved for the flow through a rotating curved rectangular duct at large pressure gradient with large rotational speed. This motivated the present study to investigate the non-linear behavior of the unsteady solutions by time-evolution calculation.

In the present study, a comprehensive numerical result is presented for fully developed bifurcation structure of two-dimensional (2D) viscous incompressible fluid flow through a rotating curved rectangular duct. Flow characteristics are investigated over a wide range of Taylor number $0 \le Tr \le 2500$ for the Dean number $Dn = 2000$. Studying the effects of rotation on the flow characteristics, caused by the combined action of centrifugal force and Coriolis force, is an important objective of the present study.

GOVERNING EQUATIONS

Consider that the flow is viscous and incompressible which is streaming through a rotating curved duct with rectangular cross section. Let 2h and 2l be the height and the width of the cross section. Figure 1 shows the coordinate system, where C is the center of the duct cross section and L is the radius of curvature of the duct. The x′ and y′ axes are taken to be in the horizontal and vertical directions respectively, and z′ is the coordinate along the center-line of the duct, i.e., the axial direction. The system rotates at a constant angular velocity T Ω around the y′ axis. It is assumed that the flow is uniform in the axial direction, and that it is driven by a constant pressure gradient G $G\left(G = -\frac{\partial P'}{\partial z'}\right)$ along the center-line of the duct, i.e. the main flow direction. Then the continuity equation and the Navier-Stokes equation, in terms of dimensional variables, are expressed as Continuity equation

$$\frac{\partial u'}{\partial r'} + \frac{\partial v}{\partial y'} + \frac{u'}{r'} = 0 \tag{1}$$

Momentum equations

$$\frac{\partial u'}{\partial t'} + u'\frac{\partial u'}{\partial r'} + v'\frac{\partial u'}{\partial y'} - \frac{w'^2}{r'} - 2\Omega_T w' = -\frac{1}{\rho}\frac{\partial P'}{\partial r'} + \upsilon\left[\frac{\partial^2 u'}{\partial r'^2} + \frac{\partial^2 u'}{\partial y'^2} + \frac{1}{r'}\frac{\partial u'}{\partial r'} - \frac{u'}{r'^2}\right], \tag{2}$$

$$\frac{\partial v'}{\partial t'} + u'\frac{\partial v'}{\partial r'} + v'\frac{\partial v'}{\partial y'} = -\frac{1}{\rho}\frac{\partial P'}{\partial y'} + \upsilon\left[\frac{\partial^2 v'}{\partial r'^2} + \frac{1}{r'}\frac{\partial v'}{\partial r'} + \frac{\partial^2 v'}{\partial y'^2}\right], \tag{3}$$

$$\frac{\partial w'}{\partial t'} + u'\frac{\partial w'}{\partial r'} + v'\frac{\partial w'}{\partial y'} + \frac{u'w'}{r'} + 2\Omega_T u' = -\frac{1}{\rho}\frac{1}{r'}\frac{\partial P'}{\partial z'} + \upsilon\left[\frac{\partial^2 w'}{\partial r'^2} + \frac{\partial^2 w'}{\partial y'^2} + \frac{1}{r'}\frac{\partial w'}{\partial r'} - \frac{w'}{r'^2}\right]. \tag{4}$$

Where r′ = L + x′ , and u′,v′ and w′ are the dimensional velocity components in the x′, y′ and z′ directions respectively. In Equations (1) to (4) the variables with prime denote the dimensional quantities. To non-dimensionalize the equations, we use the non-dimensional variables defined as

$$u = \frac{u'}{U_0},\ v = \frac{v'}{U_0},\ w = \frac{\sqrt{2\delta}}{U_o}w',\ x = \left(\frac{x'}{l} - \frac{1}{\delta}\right),\ \bar{y} = \frac{y'}{l},\ z = \frac{z'}{l},\ t = \frac{U_0}{l}t',\ \delta = \frac{l}{L},\ P = \frac{P'}{\rho U_0^2},$$

Where u,v and w are the non-dimensional velocity components in the x, y and z directions, respectively; t is the non-dimensional time, P is the non-dimensional pressure, δ is the non-dimensional curvature defined as $\delta = \frac{l}{L}$. The sectional stream function ψ is introduced as

$$u = \frac{1}{1+\delta x}\frac{\partial \psi}{\partial y}, \quad v = -\frac{1}{1+\delta x}\frac{\partial \psi}{\partial x}. \tag{5}$$

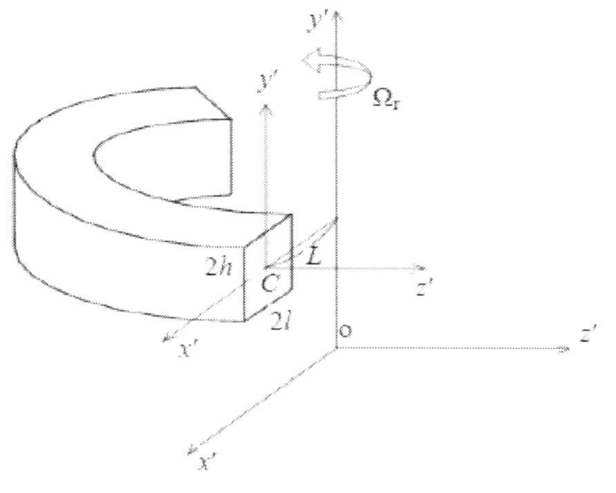

Figure 1. Rotating coordinate system

Then, the basic equations for w and ψ are expressed in terms of non-dimensional variables as

$$(1+\delta x)\frac{\partial w}{\partial t} + \frac{1}{a}\frac{\partial(w,\psi)}{\partial(x,y)} - Dn + \frac{\delta^2 w}{1+\delta x} = (1+\delta x)\Delta_2 w - \frac{\delta}{a(1+\delta x)}\frac{\partial \psi}{\partial y} w + \delta\frac{\partial w}{\partial x} - \delta Tr\frac{\partial \psi}{\partial y}, \tag{6}$$

$$\begin{aligned}\left(\Delta_2 - \frac{\delta}{1+\delta x}\frac{\partial}{\partial x}\right)\frac{\partial \psi}{\partial t} = & -\frac{1}{a(1+\delta x)}\frac{\partial(\Delta_2\psi,\psi)}{\partial(x,y)} + \frac{\delta}{a(1+\delta x)^2}\left[\frac{\partial \psi}{\partial y}\left(2\Delta_2\psi - \frac{3\delta}{1+\delta x}\frac{\partial \psi}{\partial x} + \frac{\partial^2 \psi}{\partial x^2}\right) - \frac{\partial \psi}{\partial x}\frac{\partial^2 \psi}{\partial x \partial y}\right] \\ & + \frac{\delta}{(1+\delta x)^2}\times\left[3\delta\frac{\partial^2 \psi}{\partial x^2} - \frac{3\delta^2}{1+\delta x}\frac{\partial \psi}{\partial x}\right] - \frac{2\delta}{1+\delta x}\frac{\partial}{\partial x}\Delta_2\psi + \frac{1}{a}w\frac{\partial w}{\partial y} + \Delta_2^2\psi + \frac{1}{2}Tr\frac{\partial w}{\partial y},\end{aligned} \tag{7}$$

where,

$$\Delta_2 \equiv \frac{\partial^2}{\partial x^2} + \frac{1}{a^2}\frac{\partial^2}{\partial y^2}, \quad \frac{\partial(f,g)}{\partial(x,y)} \equiv \frac{\partial f}{\partial x}\frac{\partial g}{\partial y} - \frac{\partial f}{\partial y}\frac{\partial g}{\partial x}. \tag{8}$$

The non-dimensional parameters Dn , the Dean number and Tr , the Taylor number which appear in Equations (6) and (7) are defined as:

$$Dn = \frac{Gl^3}{\mu\upsilon}\sqrt{\frac{2l}{L}}, \quad Tr = \frac{2\sqrt{2\delta}\Omega_T l^3}{\upsilon\delta} \tag{9}$$

In the present study, Tr varies while Dn, δ , a and Pr are fixed as Dn = 2000, δ = 0.1 , a = 1.5 and Pr = 7.0 (water). The rigid boundary conditions for w and ψ are used as

$$w(\pm 1, y) = w(x, \pm 1) = \psi(\pm 1, y) = \psi(x, \pm 1) = \frac{\partial \psi}{\partial x}(\pm 1, y) = \frac{\partial \psi}{\partial y}(x, \pm 1) = 0 \tag{10}$$

NUMERICAL CALCULATIONS

In order to solve Equations (6) to (7) numerically, the spectral method is used (Gottlieb and Orazag [23]). By this method the variables are expanded by using the expansion functions ( ) n φ x and ( ) n ψ x as

$$\left.\begin{aligned} \phi_n(x) &= (1-x^2)C_n(x), \\ \psi_n(x) &= (1-x^2)^2 C_n(x) \end{aligned}\right\} \tag{11}$$

Where $C_n(x) = \cos(n\cos^{-1}(x))$ is the nth order Chebyshev polynomial? w(x, y,t) and ψ (x, y,t) are expanded in terms of the expansion functions ( ) n φ x and ( ) n ψ x as

$$\left.\begin{aligned} w(x,y,t) &= \sum_{m=0}^{M}\sum_{n=0}^{N} w_{mn}(t)\phi_m(x)\phi_n(y), \\ \psi(x,y,t) &= \sum_{m=0}^{M}\sum_{n=0}^{N} \psi_{mn}(t)\psi_m(x)\psi_n(y). \end{aligned}\right\} \tag{12}$$

First, steady solutions are obtained by the Newton-Raps on iteration method. Then, in order to calculate the unsteady solutions, the Crank-Nicolson and Adams-Bash-forth methods together with the function expansion (12) and the collocation methods are applied.

FLUX THROUGH THE DUCT

The dimensional total flux Q' through the duct in the rotating coordinate system is calculated by:

$$Q' = \int_{-h}^{h}\int_{-l}^{l} w' dx' dy' = \nu dQ, \tag{13}$$

where

$$Q = \int_{-1}^{1}\int_{-1}^{1} w dx dy \tag{14}$$

Is the dimensionless total fluxes the mean axial velocity, w' is expressed as $\bar{w}' = \frac{Q\nu}{4l}$. In the present study, Q is used to denote the steady solution branches and to pursue time evolution of the unsteady solutions.

RESULTS AND DISCUSSION

Steady Solutions and Flow Patterns

We obtained four branches of symmetric/asymmetric steady solutions for the Dean number Dn = 2000 over the Taylor number $0 \leq Tr \leq 2500$, as shown in Figure 2. The four steady solution branches are named the first steady solution branch (first branch, thick solid line), the second steady solution branch (second branch, thin solid line), the third steady solution branch (third branch, thick dotted line) and the fourth steady solution branch (fourth branch, long dash line), respectively. The steady solution branches are obtained by the path continuation technique (Keller [24]) with various initial guesses as discussed by Mondal [25]. In this regard, it should be noted that Mondal et al. [13] also obtained four branches of steady solutions for the non-isothermal flow through a rotating curved square duct.

Then, we obtained secondary vortices on various branches of steady solutions, and it is found that at the same value of Tr sometimes we obtain two-vortex solution, while sometimes two- and multi-vortex solutions. It is found that the first and fourth steady solution branche consists of symmetric solutions while the second and third branches asymmetric solutions. The first branch consists of symmetric two-, four-, six-, eight- and ten-vortex solutions, the second branch is composed of asymmetric two- and four-vortex solutions, the third branch is characterized by asymmetric two- and four-vortex solutions, while the fourth branch is comprised with asymmetric two- to ten-vortex solutions. These vortices are generated due to the combined action of the centrifugal force and Coriolis force. The steady solution branches as well as the formation of secondary vortices on various branches are not shown here for brevity; however, we show some contours of secondary flow patterns and axial flow distribution at some specific values of Tr. To observe the pattern variation and development of the secondary vortices, also called the Dean vortices, contours of secondary flow patterns are shown in Figure 3 at various values of Tr for Dn = 2000. In the figures of the secondary flow, solid lines ($\psi \geq 0$) show that the secondary flow is in the counter clockwise direction while the dotted lines ($\psi < 0$) in the clockwise direction. As seen in Figure 3, the secondary flow is two-, four-, six-, eight- and ten-vortex solutions, which are obtained on different branches of steady solutions. Figure 4 shows typical contours of secondary flow patterns and axial flow distribution on the steady solution branch at Tr = 600, where it is found that the secondary flow consists of symmetric and asymmetric two- to ten-vortex solutions at Tr = 600 .

Unsteady Solutions

In order to investigate the non-linear behavior of the unsteady solutions, time-evolution calculations of the unsteady solutions are performed for Dn = 2000 over the Taylor number $0 \le Tr \le 2500$ Time evolution of Q for $Tr \ge 2030$ shows that the value of Q quickly approaches steady-state solution no matter what the initial condition we use. Then, in order to see the unsteady flow characteristics for Tr < 2030 and Tr > 2030, time

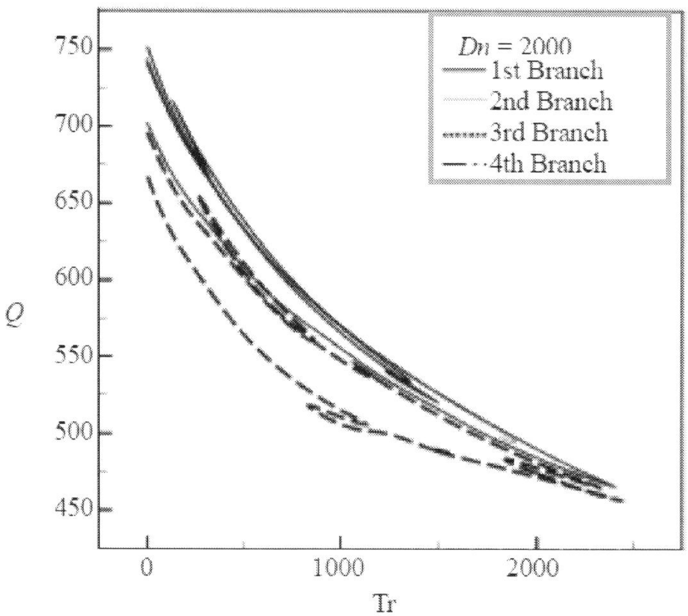

Figure 2. Solution structure of the steady solutions for $0 \le Tr \le 2500$ at Dn = 2000.

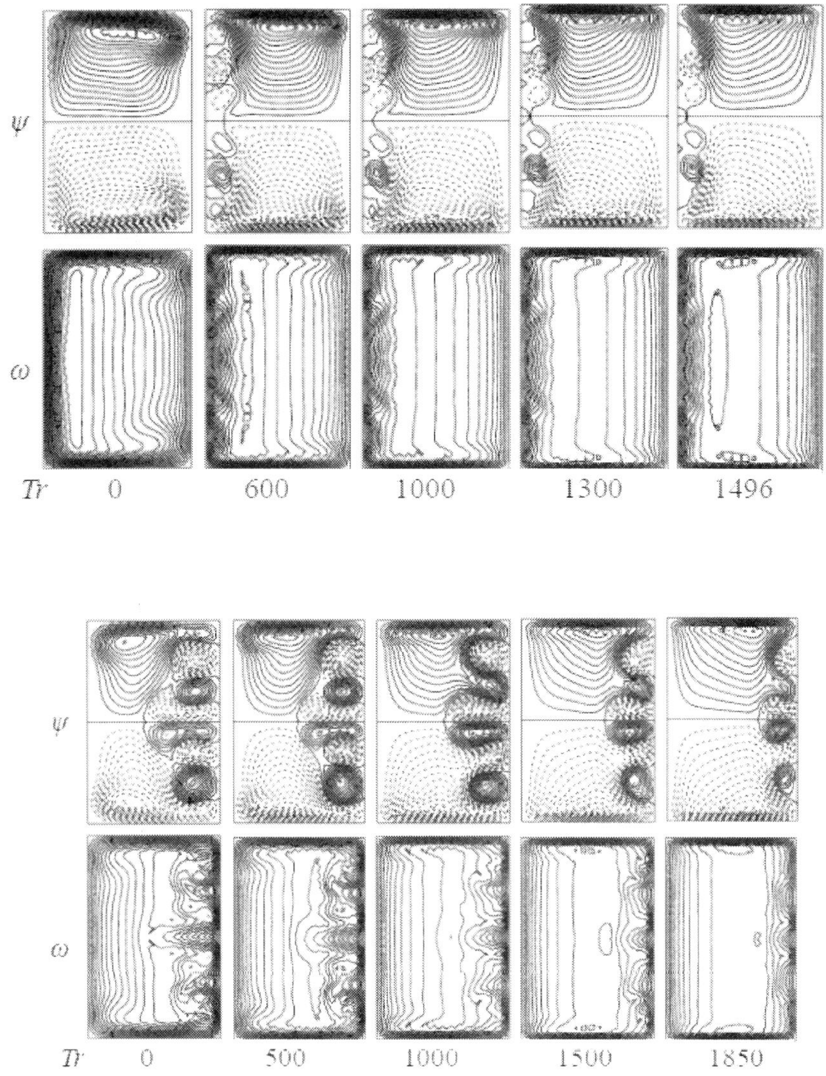

Figure 3. Streamlines of secondary flow (top) and axial flow (bottom) on the steady solution branches at various values of Tr

Evolutions of Q are then performed for Tr = 0 , 500, 1000, 1500, 1870, 1880, 1900, 1950 and 2050. Figure 5(a) shows time-evolution of Q for Tr = 0 at Dn = 2000, where it is seen that the flow oscillates irregularly, i.e. the flow is chaotic. This chaotic oscillation is well justified by drawing the power spectrum as shown in Figure 5(b), where it is seen that lots of continuous line spectra with different frequencies are available, which suggests that the unsteady flow at Tr = 0 is chaotic. Typical contours of secondary flow patterns and axial flow distribution at Tr = 0 are shown in Figure 5(c), where it is seen that the flow oscillates between asymmetric six- to eight-vortex solutions. Time-evolution of Q for Tr = 500 is shown in Figure 6(a), where it is seen that the

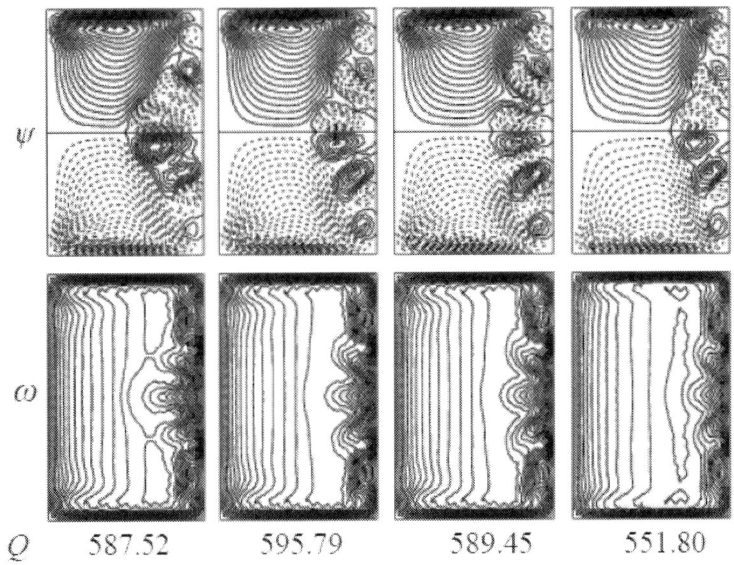

Figure4. Streamlines of secondary flow (top) and axial flow (bottom) on the fourth steady solution branch for various values of Q at Tr = 600

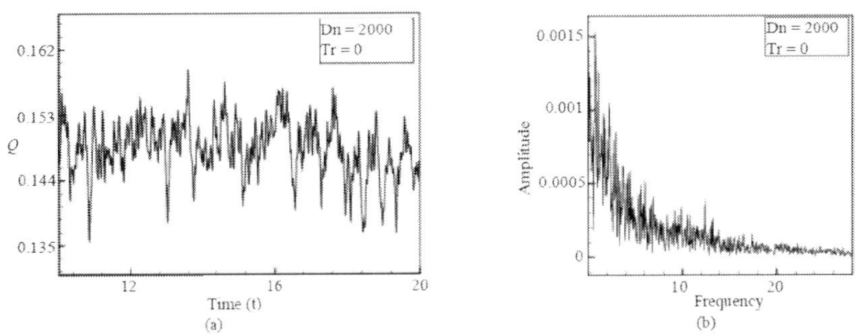

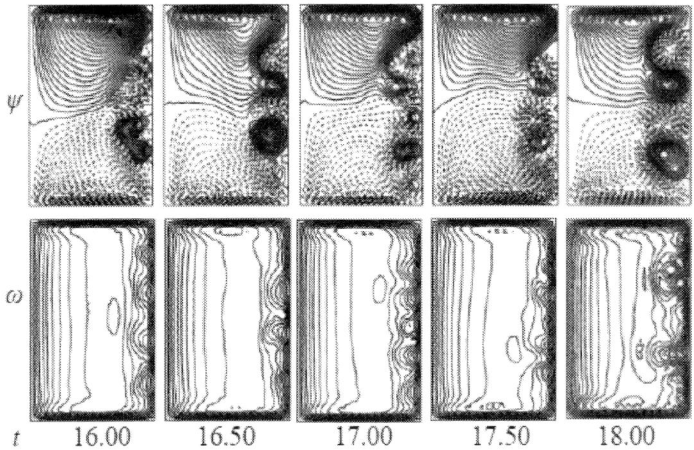

(c)

Figure 8 (a) Time evolution of Q for Dn = 2000 and Tr = 1500; (b) Power spectrum for Tr = 1500; (c) Streamlines of secondary flow (top) and axial flow (bottom) for Tr = 1500.

Figures 7(b) and **8(b)** for *Tr* = 1000 and *Tr* = 1500 respectively, where lots of continuous line spectra with different frequencies are seen, which suggests that the unsteady flow at *Tr* = 1000 and *Tr* = 1500 are chaotic. Typical contours of secondary flow patterns and axial flow distribution for *Tr* = 1000 and *Tr* = 1500 are shown in **Figures 7(c)** and **8(c)** respectively, where it is seen that the flow oscillates between asymmetric fourto ten-vortex solutions. **Figure 9(a)** shows time-evolution of *Q* for *Tr* = 1870, where it is seen that the flow is chaotic. This chaotic solution is well justified by the power spectrum of the time change of *Q* as shown in **Figure 9(b)**, where it is seen that lots of continuous line spectra with different frequencies are available, which suggests that the flow is chaotic. Secondary flow patterns and axial flow distribution for *Tr* = 1870 are shown in **Figure 9(c)**, where it is seen that the flow oscillates between asymmetric six- and eight-vortex solutions.

Then we performed time evolution of *Q* for *Tr* = 1880 as shown in **Figure 10(a)**. As seen in **Figure 10(a)**, the time-dependent solution for *Tr* = 1880 is multi-periodic. Power spectrum of the time change of *Q* for *Tr*= 1880 is also shown in **Figure 10(b)**, in which not only the line spectrum of the fundamental frequency and its harmonics but also other line spectrum and their harmonics are seen, which suggests that the flow is multi- periodic. Secondary flow patterns and axial flow distributions are shown in **Figure 10(c)**, for one period of oscillation at $5.7 \leq t \leq 5.91$ where it is seen that the unsteady flow at Tr = 1880 oscillates between asymmetric four- and six-vortex solutions.

Next, time evolution of Q for Tr = 1900 is shown in **Figure 11(a)**, where it is seen that the flow oscillates periodically. It is justified by the power spectrum as shown in **Figure 11(b)**, where the fundamental frequency and its harmonics as well as line spectra with small frequency is seen, which indicates that the oscillation presented in **Figure 11(b)** is multi-periodic, but not periodic. It is seen that the fundamental mode is higher than that of the other modes, which clearly suggests that the flow at Tr = 1900 is perfectly multi-periodic. Contours of secondary flow patterns and axial flow distributions are shown in **Figure 11(c)**. As seen in **Figure 11(c)**, the unsteady solution at Tr = 1900 oscillates in the asymmetric six-vortex solutions. Then, we investigated time-dependent solution of Q for Tr = 1950 at Dn = 2000 as shown in **Figure 12(a)**. It is found that the flow

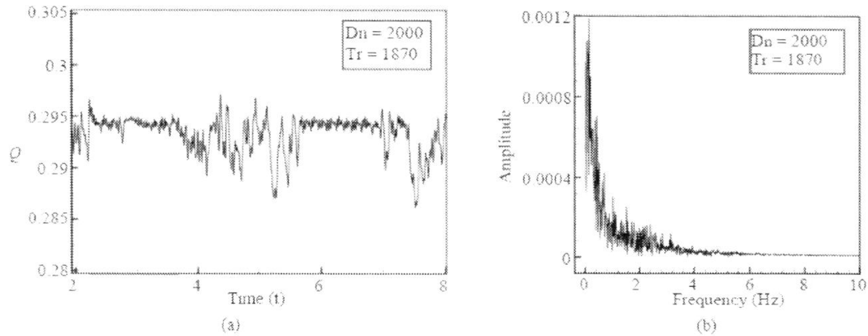

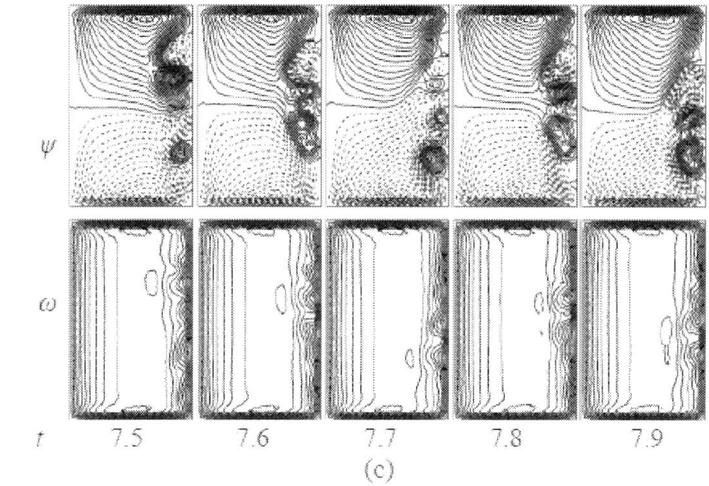

Figure 9 (a) Time evolution of Q for Dn = 2000 and Tr = 1870; (b) Power spectrum for Tr = 1870; (c) Streamlines of secondary flow (top) and axial flow (bottom) for Tr = 1870.

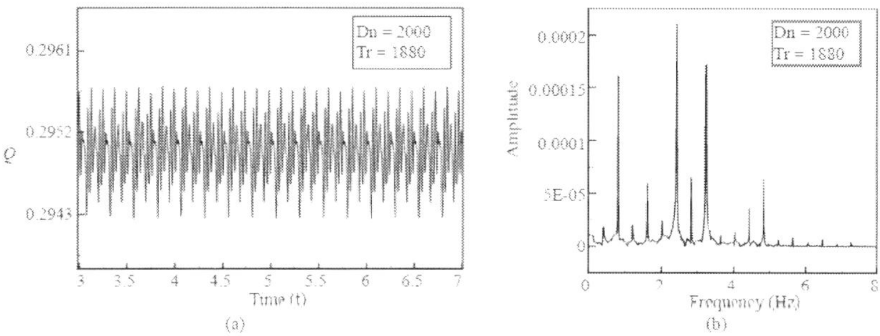

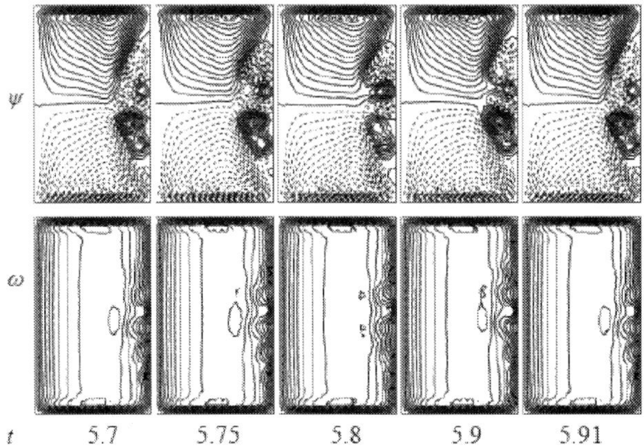

(c)

Figure10. (a) Time evolution of Q for Tr = 1880 at Dn = 2000; (b) Power spectrum for Tr = 1880; (c) Streamlines of secondary flow (top) and axial flow (bottom) for Tr = 1880.

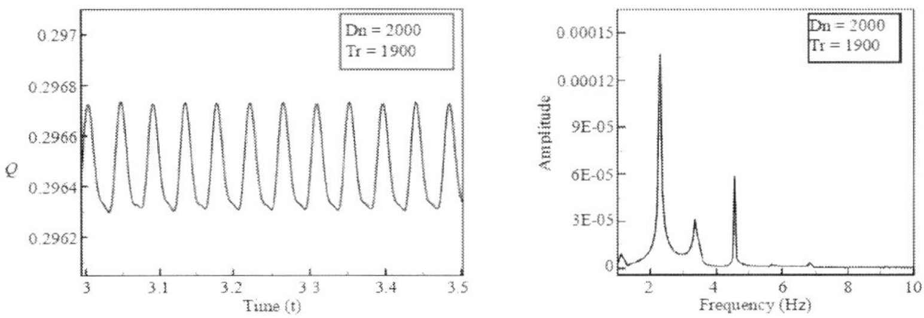

(a) (b)

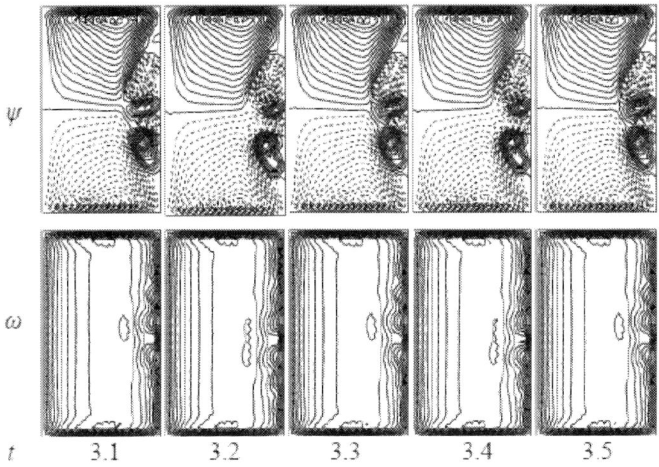

(c)

Figure 11 (a) Time evolution of Q for Tr = 1900 and Dn = 2000; (b) Power spectrum for Tr = 1900; (c) Streamlines of secondary flow (top) and axial flow (bottom) for Tr = 1900.

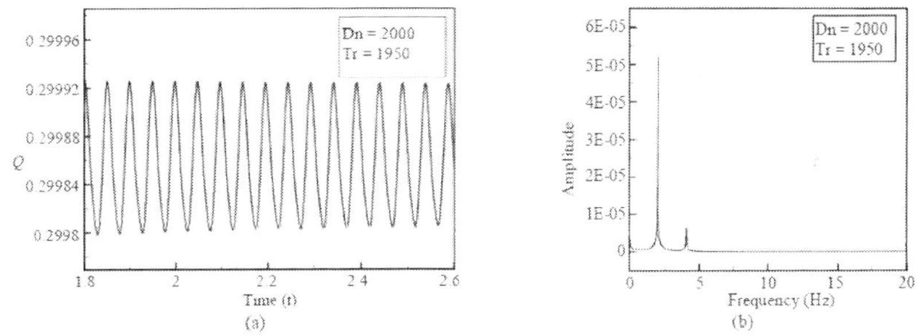

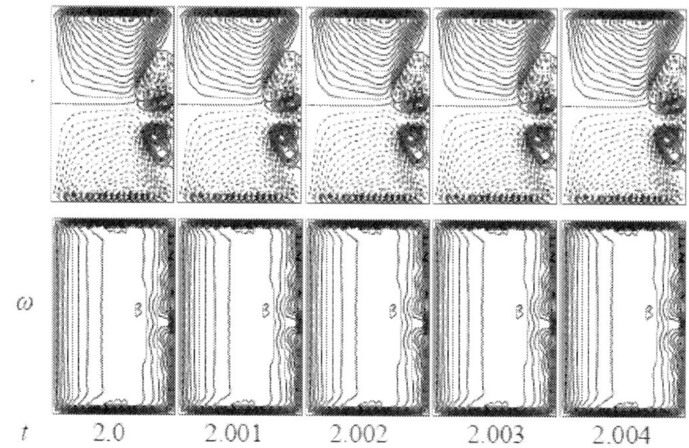

(c)

Figure 12 (a) Time evolution of Q for Tr = 1950 at Dn = 2000; (b) Power spectrum for Tr = 1950; (c) Streamlines of secondary flow (top) and axial flow (bottom) for Tr = 1950.

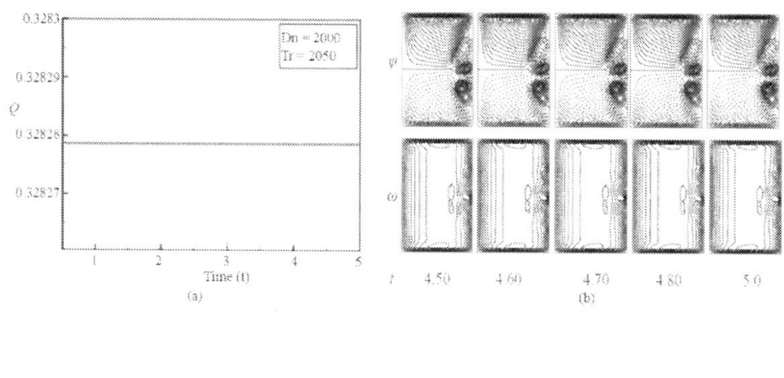

Figure 13. (a) Time evolution of Q for Dn = 2000 and Tr = 2050; (b) Streamlines of secondary flow (top) and axial flow (bottom) for Tr = 2050.

oscillates periodically, which is well justified by the power spectrum of the time evolution of Q as shown in Figure 12(b), where only the line spectrum of the fundamental frequency and its harmonics are seen, which suggests that the flow is purely periodic at Tr = 1950 . Secondary flow patterns and axial flow distribution for Tr = 1950 are then shown in Figure 12(c), where it is seen that the flow at Tr = 1950 oscillates in the asymmetric six-vortex solutions. Figure 13(a) shows time evolution of Q for Tr = 2050 at Dn = 2000. It is found that the flow does not oscillate but goes steady-state, so that the unsteady solution at Tr = 2050 is a steady-state solution. Typical contours of secondary flow patterns and axial flow distribution for Tr = 2050 are also shown in Figure 13(b), where it is seen that the unsteady flow at Tr = 2050 is an asymmetric six-vortex solution. It is noticed that axial flow is shifted near the outer wall of the duct as the rotational speed increases.

CONCLUSIONS

A numerical study on the fully developed two-dimensional flow of viscous incompressible fluid through a rotating curved rectangular duct of aspect ratio 1.5 and curvature 0.1 has been performed by using the spectral method, and covering a wide range of the Taylor number $0 \leq Tr \leq 2500$ for the Dean number Dn = 2000. We investigated flow characteristics for positive rotation of the duct, and obtained secondary flow patterns and axial flow distribution for several values of Tr . We obtained four branches of symmetric and asymmetric steady solutions with two- and multi-vortex solutions on various branches. The first branch consists of symmetric two-, four- and multi-vortex solutions. The second branch is composed of asymmetric two-, three- and four-vortex solutions; the third steady solution branch is composed of asymmetric two-, three- and four-vortex solutions, while the forth steady solution branch symmetric six-, eight- and ten-vortex solutions. Then, In order to study the non-linear behavior of the unsteady solutions, time-evolution calculations as well as power spectrum of the unsteady solutions are performed, and it is found that the chaotic flow turns into steady-state flow through periodic flow in the scenario "chaotic → multi-periodic → periodic → steady-state", if Tr is increased from zero. It is found that at no ration, the flow is chaotic but as the rotational speed increases, the chaotic flow turns into steady-state flow through multi-periodic and periodic oscillations. The reason is that combined effect of centrifugal and Coriolis force counteract each other in a nonlinear manner which results in chaotic flow to turn into steady-state flow. Performing the power spectrum of the solutions was found to be fruitful for the investigation of unsteady flow behavior more accurately. Maximum axial flow was found to be shifted near the outer wall of the duct as the rotational speed increases.

ACKNOWLEDGEMENTS

Shinichiro Yanase, one of the authors, expresses his cordial thanks to the Japan Ministry of Education, Culture, Sports, Science and Technology for the financial support through the Grant-in-Aid for Scientific Research, No. 24560196.

REFERENCES

1. Ishigaki, H. (1993) Fundamental Characteristics of Laminar Flows in a Rotating Curved Pipe. Transactions of JSME, 59, 1494-1501.
2. Ishigaki, H. (1996) Laminar Flow in Rotating Curved Pipes. Journal of Fluid Mechanics, 329, 373-388.
3. Berger, S.A., Talbot, L. and Yao, L.S. (1983) Flow in Curved Pipes. Annual Review of Fluid Mechanics, 35, 461-512.
4. Nandakumar, K. and Masliyah, J.H. (1986) Swirling Flow and Heat Transfer in Coiled and Twisted Pipes. Advanced Transport Processes, 4, 49-112
5. Winters, K.H. (1987) A Bifurcation Study of Laminar Flow in a Curved Tube of Rectangular Cross-Section. Journal of Fluid Mechanics, 180, 343-369.
6. Mondal, R.N., Kaga, Y., Hyakutake, T. and Yanase, S. (2007) Bifurcation Diagram for Two-Dimensional Steady Flow and Unsteady Solutions in a Curved Square Duct. Fluid Dynamics Research, 39, 413-446.
7. Selmi, M., Namdakumar, K. and Finlay, W.H. (1994) A Bifurcation Study of Viscous Flow through a Rotating Curved Duct. Journal of Fluid Mechanics, 262, 353-375.
8. Selmi, M. and Namdakumar, K. (1999) Bifurcation Study of the Flow through Rotating Curved Ducts. Physics of Fluids, 11, 2030-2043.
9. Yamamoto, K., Yanase, S. and Alam, M.M. (1999) Flow through a Rotating Curved Duct with Square Cross-Section. Journal of the Physical Society of Japan, 68, 1173-1184.
10. Mondal, R.N., Islam, M.R., Uddin, M.S. and Datta, A.K. (2010) Flow through a Rotating Curved Square Duct: The Case of Positive Rotation. Journal of Physical Science, 14, 145-163.
11. Wang, L.Q. and Cheng, K.C. (1996) Flow Transitions and Combined

Free and Forced Convective Heat Transfer in Rotating Curved Channels: The Case of Positive Rotation. Physics of Fluids, 8, 1553-1573.

12. Zhang, J.S., Zhang, B.Z. and Jü, J. (2001) Fluid Flow in a Rotating Curved Rectangular Duct. International Journal of Heat and Fluid Flow, 22, 583-592.
13. Mondal, R.N., Alam, M.M. and Yanase, S. (2007) Numerical Prediction of Non-Isothermal Flows through a Rotating Curved Duct with Square Cross Section. Thammasat International Journal of Science and Technology, 12, 24-43.
14. Mondal, R.N., Datta, A.K. and Mondal, B. (2011) Bifurcation Study of Thermal Flows through a Rotating Curved Square Duct. Bangladesh Journal of Scientific and Industrial Research, 48, 59-70.
15. Yamamoto, K., Wu, X., Nozaki, K. and Hayamizu, Y. (2006) Visualization of Taylor-Dean Flow in a Curved Duct of Square Cross-Section. Fluid Dynamics Research, 38, 1- [16] Nobari, M.R.H., Nousha, A. and Damangir, E. (2009) A Numerical Investigation of Flow and Heat Transfer in Rotating U-Shaped Square Ducts. International Journal of Thermal Sciences, 48, 590-601.
16. Yanase, S. and Nishiyama, K. (1988) On the Bifurcation of Laminar Flows through a Curved Rectangular Tube. Journal of the Physical Society of Japan, 57, 3790-3795.
17. Yanase, S., Kaga, Y. and Daikai, R. (2002) Laminar Flow through a Curved Rectangular Duct over a Wide Range of the Aspect Ratio. Fluid Dynamics Research, 31, [19] Yang, T. and Wang, L. (2003) Bifurcation and Stability of Forced Convection in Rotating Curved Ducts of Square Cross Section. International Journal of Heat and Mass Transfer, 46, 613-629.
18. Wang, L.Q. and Yang, T.L. (2004) Multiplicity and Stability of Convection in Curved Ducts: Review and Progress. Advances in Heat Transfer, 38, 203-256.
19. Yanase, S., Mondal, R.N. and Kaga, Y. (2005) Numerical Study of Non-Isothermal Flow with Convective Heat Transfer in a Curved Rectangular Duct. International Journal of Thermal Sciences, 44, 1047-1060.
20. Mondal, R.N., Kaga, Y., Hyakutake, T. and Yanase, S. (2006) Effects of Curvature and Convective Heat Transfer in Curved Square Duct Flows. Journal of Fluids Engineering, ASME Journals, 128, 1013-1023.
21. Gottlieb, D. and Orazag, S.A. (1977) Numerical Analysis of Spectral Methods. Society for Industrial and Applied Mathematics, Philadelphia

22. Keller, H.B. (1987) Lectures on Numerical Methods in Bifurcation Problems. Springer, Berlin
23. Mondal, R.N. (2006) Isothermal and Non-Isothermal Flows through Curved Duct with Square and Rectangular Cross- Section. Ph.D. Thesis, Okayama University, Okayama

Chapter 2

AN INITIAL INVESTIGATION OF ADJOINT-BASED UNSTRUCTURED GRID ADAPTATION FOR VORTICAL FLOW SIMULATIONS

Li Li

Aeronautics Computing Technique Research Institute (ACTRI), P.O. Box 90, Xi'an 710068, China

ABSTRACT

A computational fluid dynamics (CFDs) method utilizing unstructured grid technology has been employed to compute vortical flow around a 65° delta wing with sharp leading edge, which is specially known as the geometry of the second international vortex flow experiment (VFE-2). In VFE-2, 65° delta wings with different leading edges had been broadly investigated by experiments, which resulted in a special database for CFDs codes validation. The emphasis of this paper is to investigate the effectiveness of an adjoint-base grid adaptation method for unstructured grid in capturing concentrated vortices generated at sharp edges or flow separation lines of lifting surfaces flying at high angles of attack. Earlier experiences in vortical flow simulations had indicated that the vortex behavior is highly dependent on the local grid resolution both on body surface and space. The adjoint-based adaptation method used here is hoped to save grid points with a reasonable grid resolution for vortical flow simulations. The basic idea is to construct a new adaptive sensor in a grid adaptation process with the intent to tell where the elements should be smaller or larger by introducing an adjoint formulation to relate the estimated functional error to local residual errors of both the primal and adjoint solutions.

INTRODUCTION

In recent years, computational fluid dynamics (CFDs) has matured as a routine tool to predict complex genuine flows at designed condition. However, it is still difficult to deal with complex flows at off-design conditions where flow separations and vortices are dominated. For these flows, inappropriate grid resolution has been recognized as one of the dominant reasons for inaccurate predictions of flow features and aerodynamic forces. Actually, in general, the computational grid may become rapidly coarser as it becomes far away from the body surface.

In this case, the vortices would be highly diffused due to the numerical discretization error. A possible way to minimize the numerical dissipation of the vortices is to use a highly dense grid. But flow computations around three-dimensional complex bodies with large-scale separations and vortices are difficult within the realistic number of grid points. In order to obtain numerical results for such flows with a given level of accuracy by a limited capacity of the computer, the method for grid adaptation is very useful for practical computations, which offers the possibilities to avoid the use of overly refined grids to guarantee accuracy. The basic premise is to locally enrich the computational grids in regions which most adversely affect the accuracy of the final solution. Its emphasis is to define a proper adaptive criterion (also known as the adaptive sensor) combined with distinct methods to automatically generate an adaptive mesh.

Currently, many sensors for unstructured grid adaptation are available, for example, the wake sensor based on difference in velocity, the shock sensor based on ratio of total pressure, the vortex sensor based on creation of entropy, and the vortex sensor based on eigenvalue analysis of the velocity-gradient tensor, and so forth

[1]. These adaptation methods are often referred to as feature-based adaptation, which mainly focus on resolving discontinuities or strong gradients in the flow field. The assumption herein is that regions of larger gradients are associated with regions of larger error. The use of unstructured grid technology for CFDs simulations allows more freedom in adapting the discretization of the meshes to improve the fidelity of the simulation. Unfortunately, continuous local refinement of the dominant features of the flow does not necessarily guarantee that the global error will simultaneously be reduced. In some cases, this may even lead to incorrect results [2]. The reason is that the adaptive sensor has failed to correctly indicate the location where the elements should be larger or smaller due to errors elsewhere in the flow field. Also, since the feature-based adaptation methods are nearly without regard to an output quality (e.g., lift coefficient, drag coefficient, moment coefficient, and so on), they may distort to resolve the flow in regions with large local error that may have a minimal effect on the output.

In this paper, an adjoint-based adaptation with isotropic h-refinement for unstructured grid is introduced to catch the vortex-dominant flow. These kinds of methods are firstly proposed in finite element communities [3] and were then extended to finite volume computations, for example, [4–10]. The basic idea of the adjoint-based adaptation method is to construct an adaptive sensor from a more robust and accurate error estimation by introducing the adjoint formulation to relate the estimated functional error to local residual errors of both the primal and adjoint solutions. The adjoint solutions provide a very powerful approach to compute output error estimation as well as to systematically adapt grids to reduce spacial discretization errors. It has also been observed that the adjoint-based adaptation can greatly reduce unnecessary grid point compared with the feature-based adaptation both for 2D and 3D flows [11]. The application of the adjoint-based adaption here is hoped to economize the grid points with a reasonable grid resolution for vortical flow simulations.

The test case is a 65° swept delta wing with sharp-leading edge from the second interactional vortex experiment (VFE-2) whose experimental data have been authorized to use by German Aerospace Center (DLR), Göttingen, Germany. This test case is also from NASA Langley national transonic facility (NTF) that has been used to illustrate effectiveness of the vortex sensors developed by us [1]. Besides, inviscid transonic flows over RAE2822 airfoil and ONERA M6 wing are given to validate the adjoint-based adaptation method used here.

METHODOLOGY

Flow Solver

The flow solver used in this paper is WoF90, which is developed by ACTRI. It is a solver employing hybrid unstructured grid both for Euler and Reynolds-averaged Navier-Stokes (RANS) equations. The solver can deal with arbitrary types of unstructured elements including tetrahedrons, hexahedrons, prisms, and pyramids. It is based on an edge-based formulation and uses a node-centered finite volume technique to solve the governing equations. The control volumes are nonoverlapping and are formed by a dual grid, which is computed from the control surfaces for each edge of the primary input mesh. The equations are integrated explicitly towards steady state with Runge-Kutta time integration. The spatial discretization is either central with artificial dissipation or upwind; both approaches are second order accurate. The solver also adopts an agglomeration multigrid algorithm and an implicit residual smoothing algorithm to accelerate the convergence of a simulation.

In this current study, Euler equation is solved by WoF90 for conserved variables, $\mathbf{Q} = (\rho, \rho u, \rho v, \rho w, E)^T$, where ρ is density, u, v, w is density, is a velocity component, is total energy per unit volume. The semidiscretization of Euler equation in space using finite volume technique leads to the following ordinary differential equation (ODE):

$$V_k \frac{d\mathbf{Q}}{dt} + \mathbf{R}(\mathbf{Q}) = 0, \tag{1}$$

where R is the flow equation residual to steady-state solution and V_k is volume of a control volume.

Adjoint Solver

Let f is the objective output function we are interested in. Often we can choose lift coefficient, drag coefficient, moment coefficient, or their combinations, and so on, as the objective output quality. The adjoint variable λ is defined as the linearized effect of a flow residual source term on the output function [10, 11]:

$$\lambda = \frac{\partial f}{\partial \mathbf{R}}. \tag{2}$$

It can be found by solving a set of linear equations, which is derived with application of the chain rule for differential

$$\left(\frac{\partial \mathbf{R}}{\partial \mathbf{Q}}\right)^T \lambda = \left(\frac{\partial f}{\partial \mathbf{Q}}\right)^T. \tag{3}$$

After the flow solution is known, this set of linear equations can be solved much like the flow solver by adding a pseudotime item in order to result in a more robust adjoint solver [7, 12]

$$\frac{\partial \lambda}{\partial t} + \mathbf{R}_\lambda(\lambda) = 0, \tag{4}$$

where R_λ (λ) is the adjoint equation residual

$$\mathbf{R}_\lambda(\lambda) = \left(\frac{\partial \mathbf{R}}{\partial \mathbf{Q}}\right)^T \lambda - \left(\frac{\partial f}{\partial \mathbf{Q}}\right)^T. \tag{5}$$

This equation has a form similar to the flow equation so that it can be integrated toward steady solution with the same Rung-Kutta time integration as the solver. An inviscid adjoint solver has been implemented in WoF90.

Adaptive Sensor Based on Error Estimation with Adjoint Solution

The adjoint solutions can provide an efficient way to estimate error of the objective output function.*f* following the derivation in [7–9], we give a brief derivation here. Firstly, let us assume Q° is a solution on the initial coarser mesh, which can be defined as a perturbation to a solution on a much finer mesh Q* as

$$\mathbf{Q}^0 = \mathbf{Q}^* + \delta \mathbf{Q}, \tag{6}$$

where the fine-mesh solution Q* and the δ Q error are not known in practice. The resulting errors in the output function and the flow equation residual are, respectively,

$$\begin{aligned} \delta f &= f\left(\mathbf{Q}^0\right) - f\left(\mathbf{Q}^*\right), \\ \delta \mathbf{R} &= \mathbf{R}\left(\mathbf{Q}^0\right) - \mathbf{R}\left(\mathbf{Q}^*\right) = \mathbf{R}\left(\mathbf{Q}^0\right). \end{aligned} \tag{7}$$

Here, the residual on the much finer mesh is assumed to be zero, that is,R(Q*) = 0 : Linearizing about the fine-mesh solution yields,

$$\delta f \cong \frac{\partial f}{\partial \mathbf{Q}^*}\delta \mathbf{Q}, \tag{8}$$

$$\delta \mathbf{R} \cong \frac{\partial \mathbf{R}}{\partial \mathbf{Q}^*}\delta \mathbf{Q}. \tag{9}$$

From (8) and by introducing the adjoint variable from (3), we can give an alternative formula for the error in the output function as

$$\delta f \cong (\lambda^*)^T \frac{\partial \mathbf{R}}{\partial \mathbf{Q}^*}\delta \mathbf{Q} = (\lambda^*)^T \delta \mathbf{R} = (\lambda^*)^T \mathbf{R}(\mathbf{Q}^0), \tag{10}$$

where$\lambda^*$is the adjoint solution on the much finer mesh, which is often undesirable. Instead, it is assumed an approximate adjoint solution $\lambda^0$is available as a substitute. In a notation of the error of adjoint solution $\delta\lambda = \lambda^0 - \lambda^*$, equation (10) can now be rewritten as

$$\begin{aligned} \delta f &\cong (\lambda^0)^T \mathbf{R}(\mathbf{Q}^0) + (\lambda^* - \lambda^0)^T \mathbf{R}(\mathbf{Q}^0) \\ &= (\lambda^0)^T \mathbf{R}(\mathbf{Q}^0) - (\delta\lambda)^T \delta \mathbf{R} \\ &= (\lambda^0)^T \mathbf{R}(\mathbf{Q}^0) - (\delta\lambda)^T \frac{\partial \mathbf{R}}{\partial \mathbf{Q}^*}\delta \mathbf{Q} \\ &= (\lambda^0)^T \mathbf{R}(\mathbf{Q}^0) + (\mathbf{R}_\lambda^*(\lambda^0))^T (\mathbf{Q}^* - \mathbf{Q}^0), \end{aligned} \tag{11}$$

Where $R^*_\lambda(\lambda)$is the adjoint equation residual on the much finer mesh, and $R^*_\lambda(\lambda)=0$ is assumed.

As shown in (11), the error in the output function can be classified two classes: a computable correction $(\lambda^0)^T \mathbf{R}(\mathbf{Q}^0)$that can be evaluated with the solution on current coarse mesh and a remaining error that generally cannot be evaluated without solving for qualities on the much finer mesh, that is,

$$\begin{aligned} \varepsilon &\cong (\lambda^* - \lambda^0)^T \mathbf{R}(\mathbf{Q}^0) \\ &\cong (\mathbf{R}_\lambda^*(\lambda^0))^T (\mathbf{Q}^* - \mathbf{Q}^0). \end{aligned} \tag{12}$$

The average of the absolute values of the two remaining errors is defined as the adjoint-based adaptive sensor

$$\varepsilon = \frac{\left|(\lambda^* - \lambda^0)^T \mathbf{R}(\mathbf{Q}^0)\right| + \left|(\mathbf{R}_\lambda^*(\lambda^0))^T (\mathbf{Q}^* - \mathbf{Q}^0)\right|}{2}, \tag{13}$$

where, in practice, both the adjoint solution error $\lambda^* - \lambda$ and the flow solution error $Q^* - Q$ would be replaced by interpolation errors

$$\lambda^* - \lambda^0 = \lambda^h - \lambda^0, \qquad \mathbf{Q}^* - \mathbf{Q}^0 = \mathbf{Q}^h - \mathbf{Q}^0, \tag{14}$$

and λ^h, Q^h are obtained by interpolating $\lambda^0, \mathbf{Q}^0$ into a globally refined mesh.

GRID ADAPTATION AND SMOOTHING

A general introduction to the flow chart of grid adaptation can be found in [1, 13]. In the current study, the local h-refinement technique is used due to its robustness and efficiency, which divides existing elements in the mesh according to initial solution (also see examples in [14, 15]). For an inviscid calculation, the unstructured grid only has elements of the tetrahedrons in three dimension (3D) or of the triangles in two dimension (2D). These elements are refined isotropically employing the binary division according to marked edges. It should be pointed out that since the binary division will bring the so-called "hanging node" and unavoidable elements differ from tetrahedrons and triangles, the method to eliminate these nodes or elements is also used here [16]. Besides, after adaptation the mesh may have a worse quality (e.g., the aspect ratio) for some of elements, while they may significantly impact the accuracy and convergence of a solution. We recommend to import the adapted mesh into a general grid generator (e.g., ICEM-CFDs, etc.) for smoothing the mesh in order to improve the global mesh quality.

To use the adjoint-based adaptive sensor defined in (13) and (14) to indicate where to be refined, a globally h-refinement mesh is firstly created by Octree division, in which a new node is inserted at the midpoint of each existing edge, and then both solutions of the flow equation Q^h and the adjoint solution λ^h on the much finer mesh are evaluate by piecewise linear interpolation. The local adjoint-based adaptive sensor now can be evaluated from (13) by exploring all nodes l(k) of the much finer mesh within each element

$$\varepsilon_k = \frac{1}{2}\sum_{l(k)} \left\{ \left| \left(\lambda^h - \lambda^0\right)^T \mathbf{R}\left(\mathbf{Q}^0\right) \right| + \left| \left(\mathbf{R}_\lambda^*\left(\lambda^0\right)\right)^T \left(\mathbf{Q}^h - \mathbf{Q}^0\right) \right| \right\} \tag{15}$$

The criterion for refinement is given as follows: given a user-specified desired error ε, let $\eta_k = N \cdot \varepsilon_k / \varepsilon$, where N is the total number of elements in current mesh, and judge if $\eta_k > 1$ is satisfied. In this paper, a desired error ε = 0 is used for all cases.

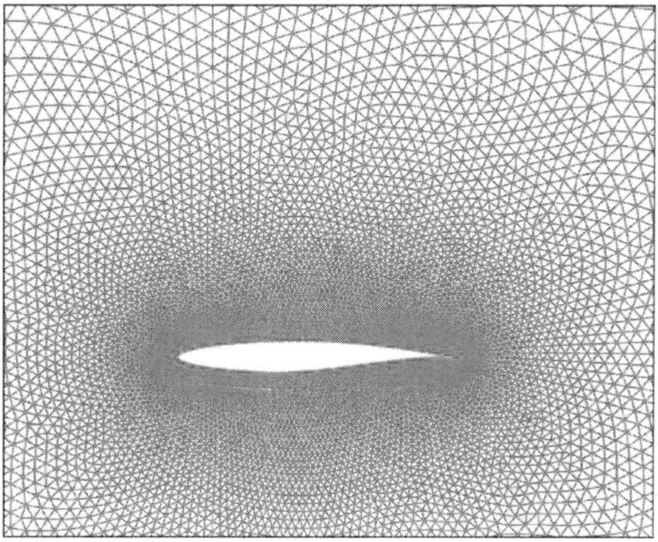

Figure 1: Initial (unadapted) RAE2822 airfoil mesh.

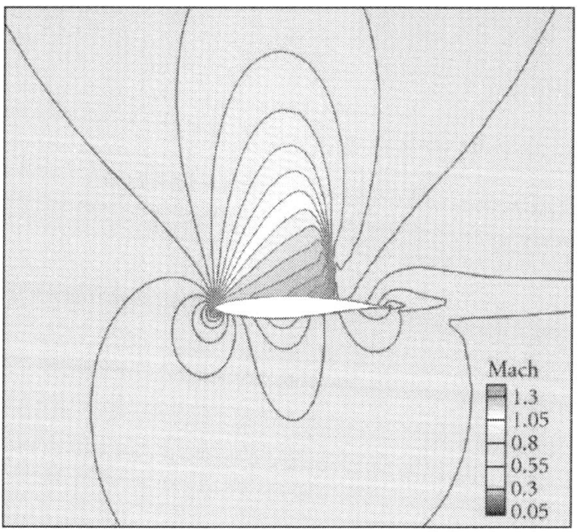

Figure 2: Initial Mach contours for RAE2822 airfoil at $M_\infty = 0.734$ and $\alpha = 2.79^\circ$.

To apply adjoint-based adaptation method, we choose the lift coefficient as the objective output function for adjoint solutions. The final adapted mesh using adjoint-based sensor and isotropic adaptation with h-refinement is shown in Figure 3, which contains a total of 20,763 nodes and 41,141 triangles after mesh adaptation. It is clearly shown that a large portion of new nodes have been appended to leading edge, upper surface and shock regions. This finding also fits the primary feature of the flow very well. The initial mesh has been sufficiently refined in high-difference regions in flow velocity.

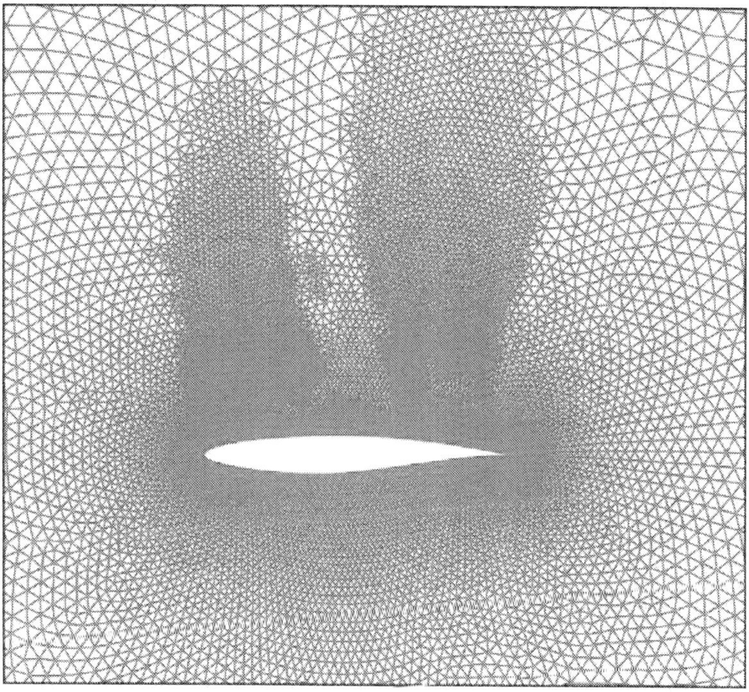

Figure 3: Adapted RAE2822 airfoil mesh.

Validation for 3D Case

Inviscid transonic flow over ONERA M6 wing is chosen to validate the rationality of adjoint-based adaptive method for three-dimensional test cases. This configuration is simulated at the flow condition at

$M_\infty = 0.8395$ and $\alpha = 3.06^\circ$. The initial mesh contains a total of 52,144 nodes and 296,517 tetrahedrons, where there are 5,459 boundary nodes on the wing surface. The initial wing surface mesh is shown in Figure 4, which has a coarse spacing, especially at the trailing edge. It is designed with the intention to resolve the surface curvature of the leading edge and the wingtip, but it has no particular considerations of flow features. For the flow solution, the central scheme is used for spatial discretization, and 3 multigrid levels with W-cycle are used for acceleration.

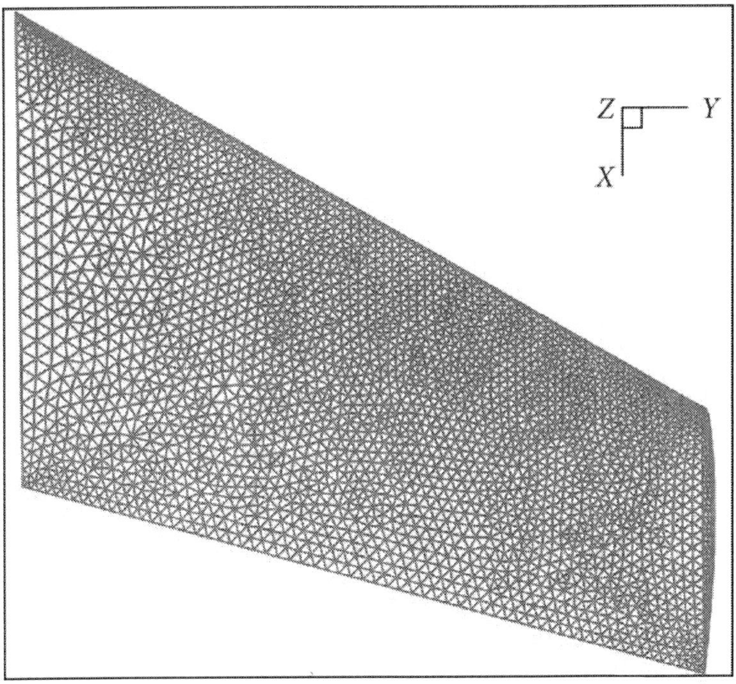

Figure 4: Initial (unadapted) surface mesh of ONERA M6 wing.

Again, we choose the lift coefficient as the objective output function for adjoint solutions. After grid adaptation using adjoint-based sensor and isotropic adaptation with h-refinement, the adapted mesh contains a total of 128,386 nodes and 740,366 tetrahedrons, where there are 8,041 boundary nodes on the wing surface. The adapted wing surface mesh is shown in Figure 5. A dense region with distinct lambda structure is found in the mesh after adaptation, which just indicates the shock region of M6 wing under current circumstance. It is also found that sufficient nodes have been enriched both at the leading edge and the trailing edge.

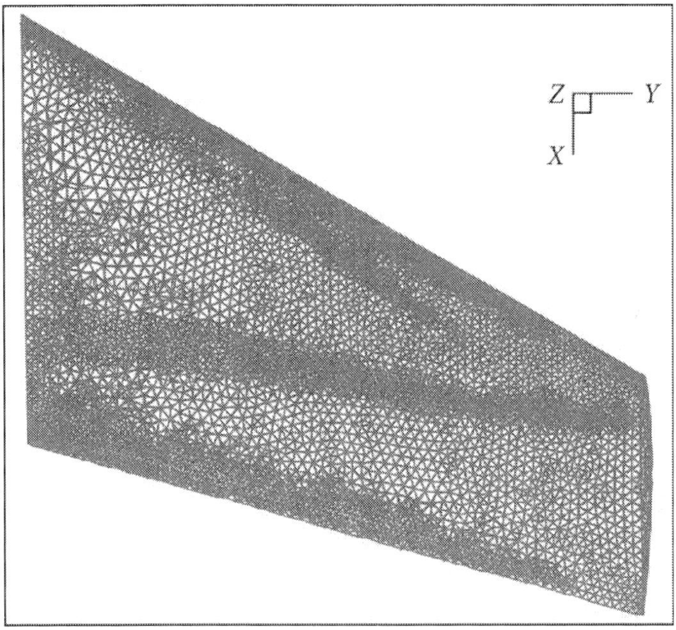

Figure 5: Adapted surface mesh of ONERA M6 wing.

The Mach contours on the adapted upper wing surface are shown in Figure 6. A lambda shock structure is clearly visible in these Mach contours, which also indicates the flow physics of M6 wing under the current flow condition. Figure 7 further gives the comparisons of the surface pressure coefficient (noted by C_p) at different typical location along the spanwise direction. The experimental data are compared with the original results and the adapted results for both upper and lower surface of the wing. It is evident that the surface pressure distribution has been successfully improved by the grid refinements with current adjoint-based adaptation method.

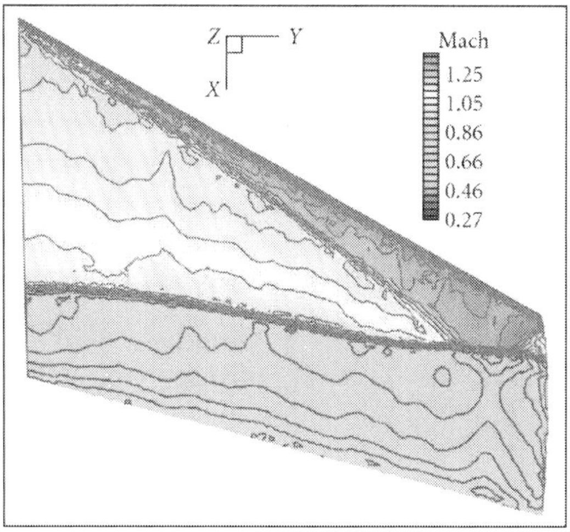

Figure 6: Mach contours on adapted surface of ONERA M6 wing at $M_\infty = 0.8395$ and $\alpha = 3.06°$.

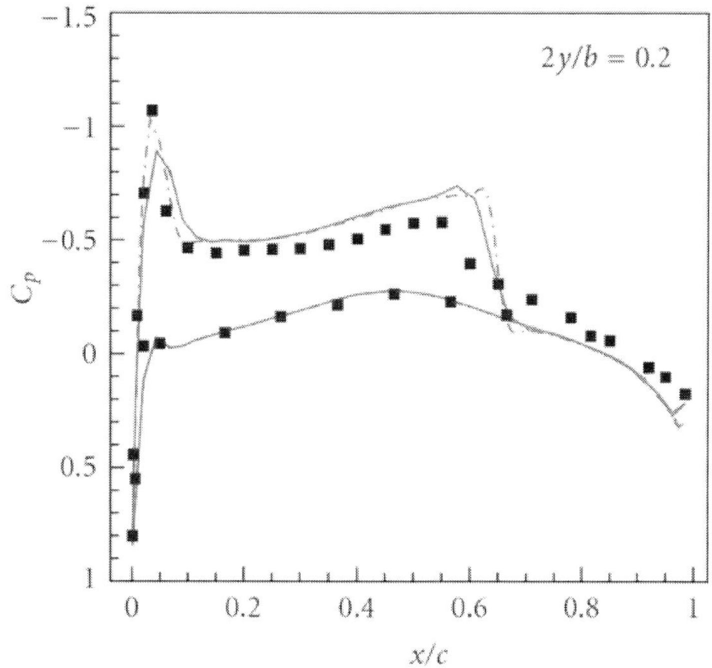

(a)

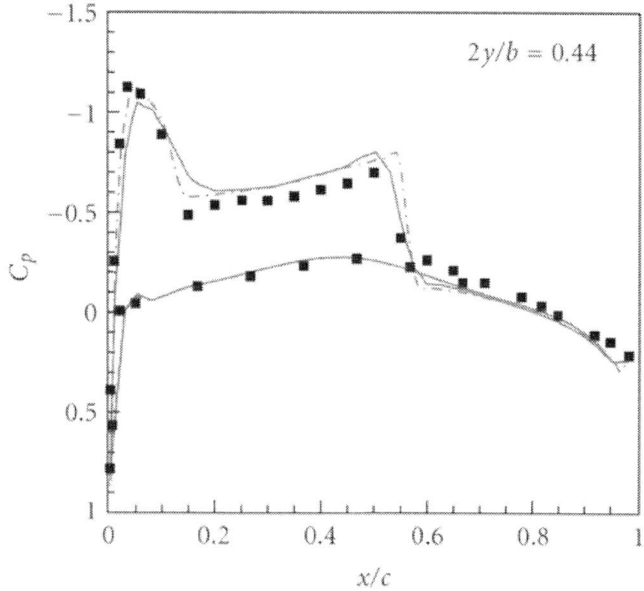
2y/b = 0.44
C_p
x/c
-1.5
-1
-0.5
0
0.5
1
0
0.2
0.4
0.6
0.8
1

(b)

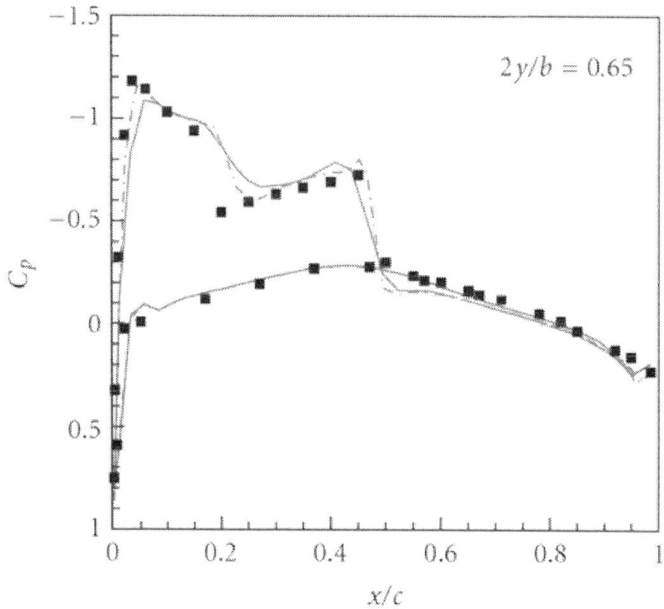
2y/b = 0.65
C_p
x/c
-1.5
-1
-0.5
0
0.5
1
0
0.2
0.4
0.6
0.8
1

(c)

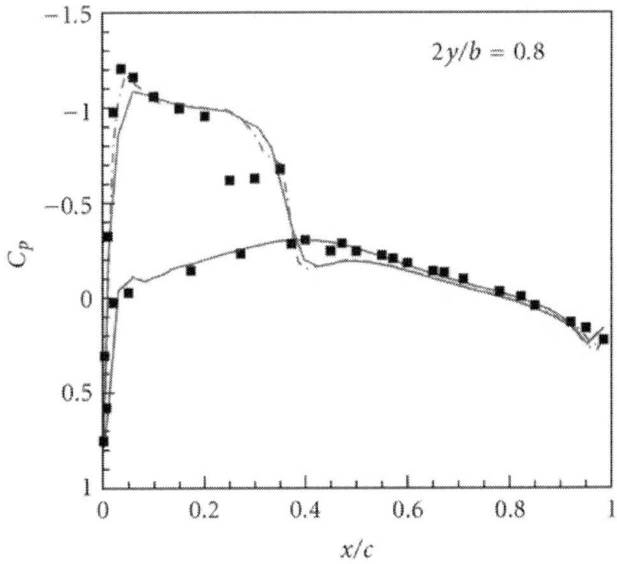
2y/b = 0.8
−1.5
−1
−0.5
0
0.5
1
C_p
0
0.2
0.4
0.6
0.8
1
x/c

(d)

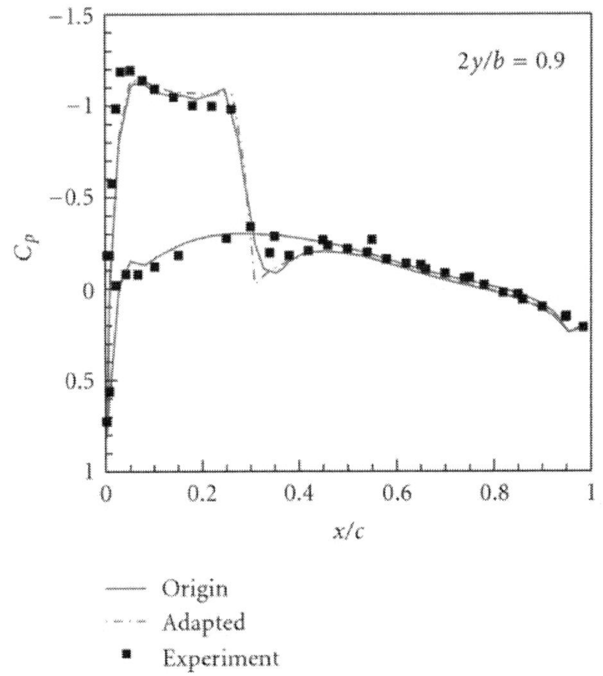
2y/b = 0.9
−1.5
−1
−0.5
0
0.5
1
C_p
0
0.2
0.4
0.6
0.8
1
x/c
Origin
Adapted
Experiment

(e)

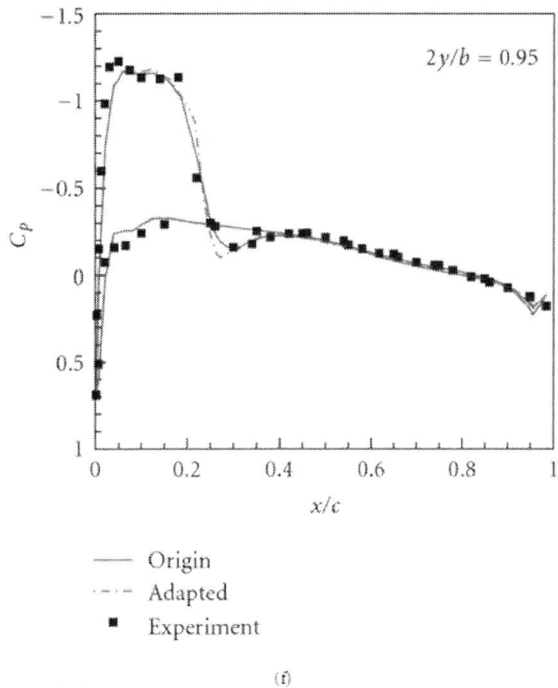

Figure 7: Comparisons of spanwise pressure coefficient distributions for ONERA M6 wing between before and after adaptation.

Application to Complex Vortical Flows of Delta Wing

As mentioned above, the test case of 65^0 delta wing with sharp leading edge is from VFE-2 as well as from NASA NTF. The emphasis of this test is to validate the vortex-capturing capability of the adjoint-based adaptation method. It is worth mentioning that while it is hard to accurately predict this kind of complex vortex flows by Euler calculation, the knowledge of vortex location obtained with grid adaptation from the inviscid solution may greatly help viscous simulations [17].

The detailed description of the model can be found in [18]. The experiment was conducted both in the NASA NTF [18] and in the transonic wind tunnel at Göttingen (TWG) of German Dutch wind tunnels (DNW) [19]. In the current study, all computations were performed on a semispan model at the flow condition at

$M_\infty = 0.4$ and $\alpha = 20.3^\circ$. For the flow equation solution, the central scheme is used for spatial discretization and 3 multigrid levels with W-cycle is used for acceleration.

The initial mesh contains a total of 64,835 nodes and 365,681 tetrahedrons, where there are 9,464 boundary nodes on the wing surface. No attempt has been made to cluster extra grid points in the field at the vortex location. The initial surface mesh on the upper surface is shown in the upper portion of Figure 8. A fine grid resolution has been prescribed at the edges of the wing, whereas the mesh on the flat portion of the wing is nearly uniform. The corresponding flow solution on this mesh is displayed in the lower portion of Figure 8, where the distribution of pressure coefficient is represented by variation of colors. The footprint of a leading-edge vortex on the surface is clearly found from the low-pressure region.

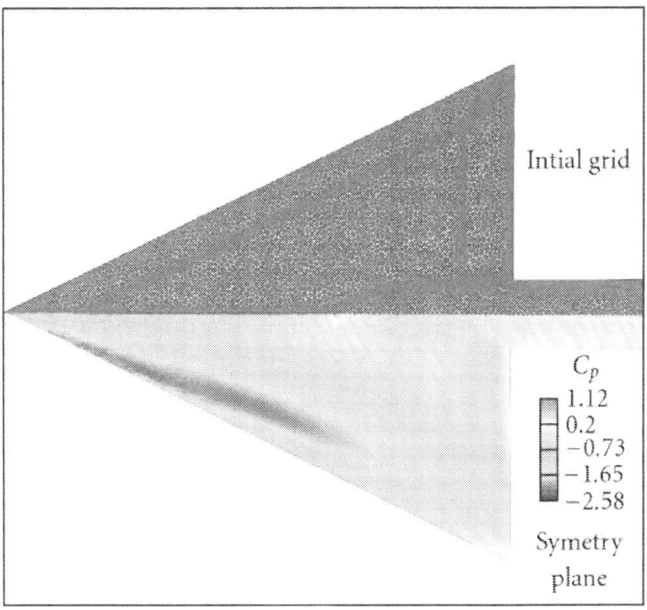

Figure 8: Initial (unadapted) upper surface mesh and Euler inviscid solution on the sharp leading-edge delta wing at $M_\infty = 0.4, \alpha = 20.3^\circ$.

To apply the adjoint-based adaptation, we also choose the lift coefficient as the output function. After the application of grid adaptation, the adapted mesh contains a total of 139,682 nodes and 766,948 tetrahedrons, where there are 13,333 nodes boundary nodes on the wing surface. The adapted upper surface mesh and the corresponding Euler solution on the adapted mesh are shown in Figure 9. It is evident that a large portion of new nodes have been enriched on the leading edge, the trailing edge, and the concentrated primary vortex regions such that the footprint of a leading-edge vortex on the surface becomes clearer. Furthermore, the primary vortex appears to be shifted slightly towards the wing root as compared with unadapted solution. In Figure 10, two typical sections at $x/c = 0.4$ and $x/c = 0.6$ were chosen for grid resolution comparison before and after mesh adaptation. Obviously, the mesh has been refined in the vortex region above the upper wing surface, while remaining unchanged under the lower wing surface. It can be also noticed that the grid density of refined regions varies with movement of the vortex core. However, it is also been found that, unlike using the adaption based on some elaborated vortex sensors [1], the spatially refined elements from adjoint-based adaptation were not over concentrated along the vortex core.

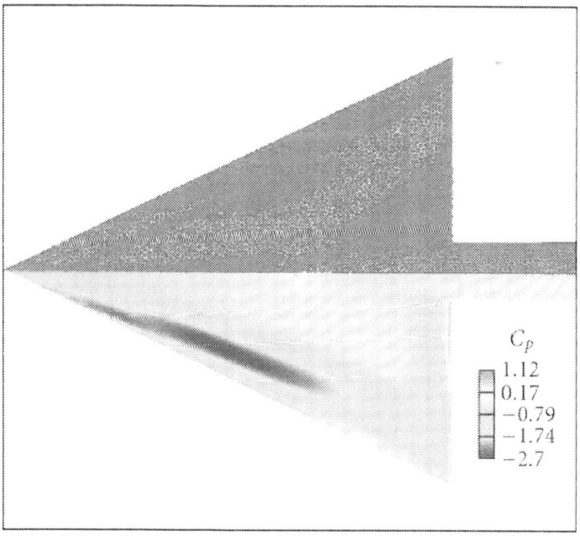

Figure 9: Adapted upper surface mesh and Euler inviscid solution on the sharp leading-edge delta wing at .

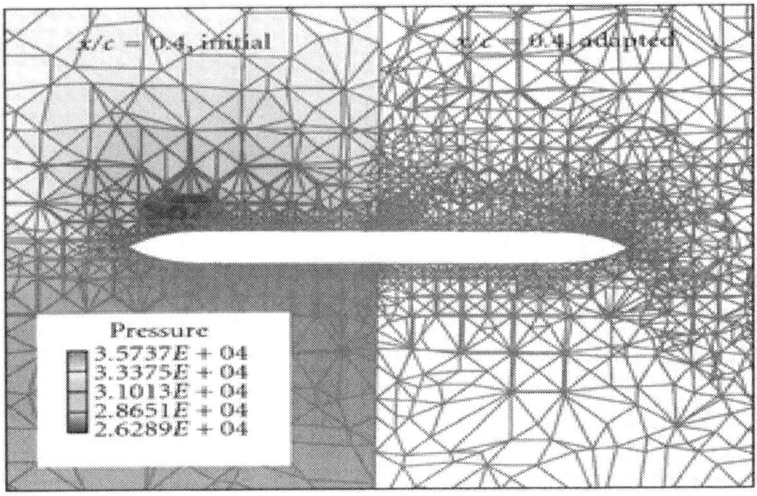

(a)

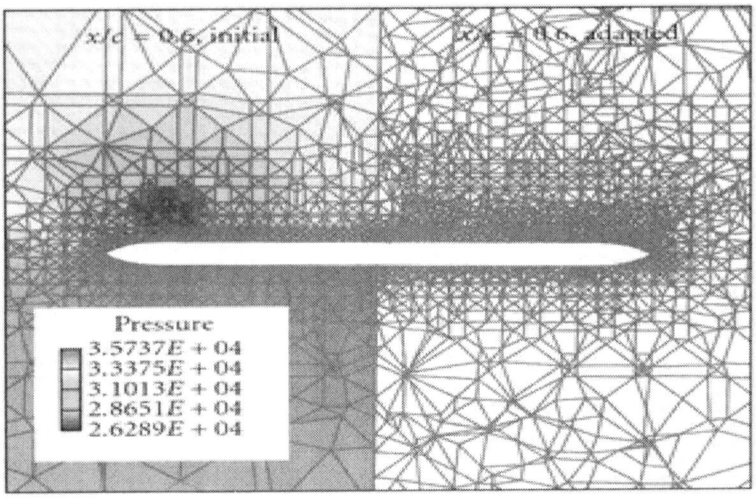

(b)

Figure 10: Comparisons of the grid resolution in space from two typical sections at $x/c = 0.4$ and $x/c = 0.$ for the sharp leading-edge delta wing.

In Figure 11, the spanwise surface pressure coefficient distributions at different chordwise positions ($x/c = 0.2, 0.4, 0.6,$, and 0.8, resp.) were compared between before and after mesh adaptation. In these figures, experimental data both from NASA and DLR are depicted. The improvement for prediction of the vortex flows is obvious by using the adjoint-based adaptation and h-refinement. The unadapted Euler simulation has produced a poor solution with underpredicted vortex in the upstream stations due to uniform grid spacing, especially; it has caused a severe shift of the peak value of C_p on the upper surface to the wing tip. After adaptation, things have been changed greatly. At all chordwise positions except $x/c = 0.2$, it appears to give a more reasonable result. At position, similarly poor results are obtained with initial and adapted meshes. It may be said that the two meshes have the same insufficient grid resolution around this section (compare Figure 8 and Figure 9). This deficiency may be ameliorated by adapting the new mesh or by altering the desired error level used in the adjoint-based adaptation.

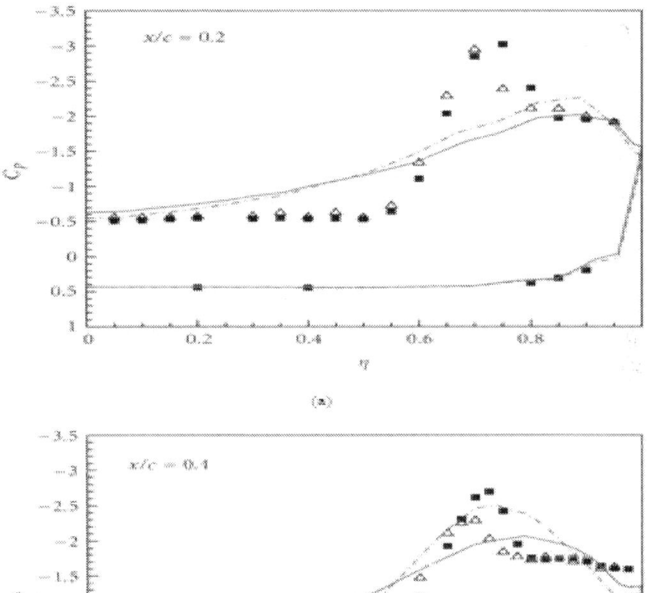

(a)

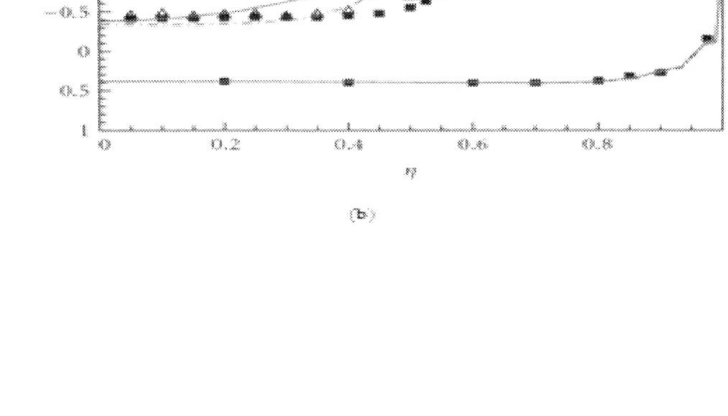

(b)

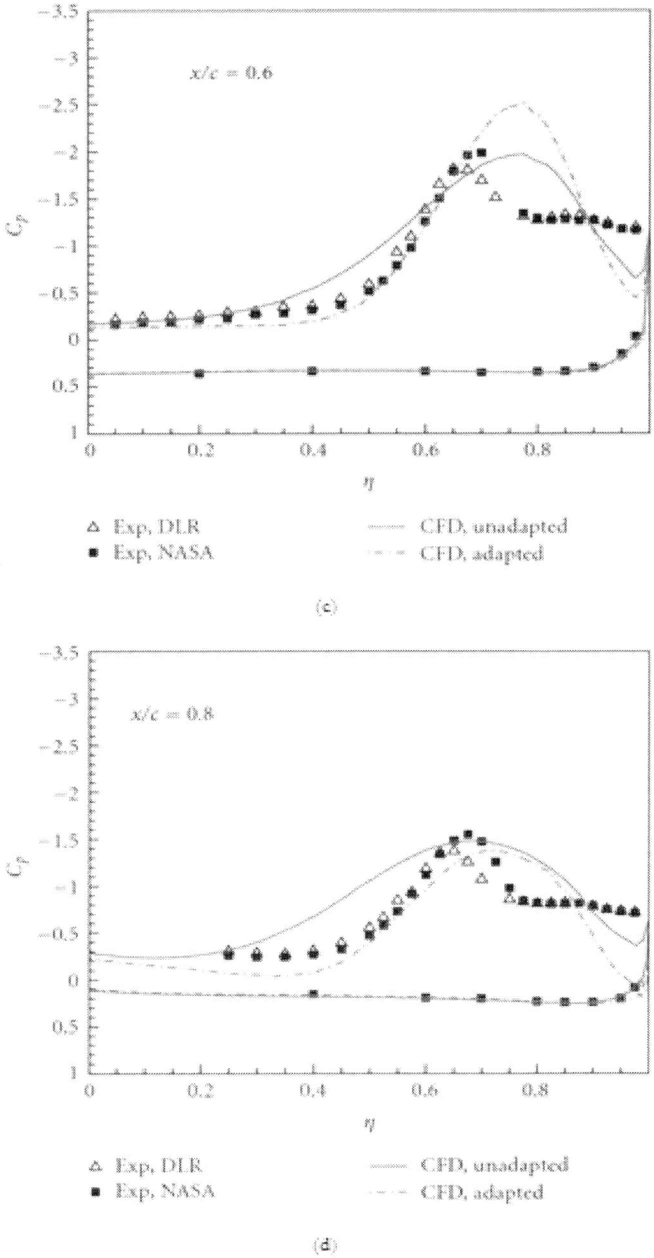

Figure 11: Comparisons of spanwise pressure coefficient distributions for delta wing with sharp leading edge between before and after mesh adaptation.

Conclusions

An adjoint-based adaptation on unstructured grid has been implemented and is applied to simulate the complex vortex flow of a 65^0 delta wing with sharp leading edge at high angle of attack. Firstly, an adaptive sensor for grid adaptation is constructed from an accurate error estimation to output function with adjoint solution, which combines residuals of the flow equation and the adjoint equation. Then, the rationality and capability of the adjoint-based adaptation is validated by choosing the transonic flow over RAE2822 airfoil as 2D case and the transonic flow over ONERA M6 wing as 3D case. Reasonable results have been obtained. After that, results from initial numerical investigation of 65^0 delta wing are given. We can conclude that although the spatial refinements from adjoint-based adaptation were not over concentrated along the vortex core of the flow, the results were greatly improved. The greatest advantage of an adjoint-based adaptation is that it easily reaches an improved and convergent grid resolution and saves unnecessary nodes which have a minimal effect on the desired output function.

It should be pointing out that although all simulations in this study are limited to inviscid Euler computations, the adjoint-based adaptive method itself can be extended to viscous calculations by developing a viscous adjoint solver and using anisotropic refinement for grid within boundary layers [9]. This work is currently under way at ACTRI.

Acknowledgments

This work is supported by the State Key Development Program of Basic Research of China (973) under Grant no. 2009CB723804. The authors would like to thank the German Aerospace center (DLR) for proving and authorizing to use the experimental data of delta wing in VFE-2 project. Mr. YANG Zhenhu, one of our colleagues at ACTRI, is also appreciated due to his part of work on adjoint-based mesh adaptive method

REFERENCES

1. B. Wen, Q. Zheng, and L. Li, "Recent efforts to establish adaptive hybrid grid computing capability at ACTRI," Journal of Computational Fluid Dynamics, vol. 7, no. 4, pp. 438–449, 2007.
2. G. P. Warren, J. T. Anderson, J. T. Thomas, and S. L. Krist, "Grid convergence for adapative methods," AIAA Paper 1991-1592, 1991.
3. P. Monk and E. Süli, "The adaptive computation of far-field patterns by a posteriori error estimation of linear functionals," SIAM Journal on Numerical Analysis, vol. 8, pp. 251–274, 1998.
4. D. A. Venditti and D. L. Darmofal, "Adjoint error estimation and grid adaptation for functional outputs: application to quasi-one-dimensional flow," Journal of Computational Physics, vol. 164, no. 1, pp. 204–227, 2000. View at Publisher · View at Google Scholar · View at MathSciNet · View at Scopus
5. J. D. Müller and M. B. Giles, "Solution adaptive mesh refinement using adjoint error analysis," AIAA Paper 2001-2550, 2001.
6. D. A. Venditti and D. L. Darmofal, "Grid adaptation for functional outputs: application to two-dimensional inviscid flows," Journal of Computational Physics, vol. 176, no. 1, pp. 40–69, 2002. View at Publisher · View at Google Scholar · View at MathSciNet · View at Scopus
7. D. A. Venditti, Grid adaptation for functional outputs of compressible flow simulations, Ph.D. thesis, Massachusetts Institute of Technology, Cambridge, Mass, USA, 2002.
8. M. A. Park, "Adjoint-based, three-dimensional error prediction and grid adaptation," AIAA Paper 2002-3286, 2002.
9. M. A. Park, "Three-dimensional turbulent RANS adjoint-based error correction," AIAA Paper 2003-3849, 2003.
10. E. M. Lee-Rausch, M. A. Park, W. T. Jones, D. P. Hammond, and E. J. Nielsen, "Application of parallel adjoint-based error estimation and anisotropic grid adaptation for three-dimensional aerospace configurations," in Proceedings of the 23rd AIAA Applied Aerodynamics Conference, vol. 1, pp. 511–533, Toronto, Canada, June 2005.
11. E. J. Nielsen, "Adjoint-based algorithms for adaptation and design optimization on unstructured grids," in Proceedings of the 3rd East-West High-Speed Flow-field Conference, Beijing, China, October 2005.
12. E. J. Nielsen, J. Lu, M. A. Park, and D. L. Darmofal, "An exact dual discrete adjoint solution method for turbulent flows on unstructured grids," AIAA Paper 2003-0273, 2003.

13. S. Z. Pirzadeh, "An adaptive unstructured grid method by grid subdivision, local remeshing, and grid movement," Tech. Rep. AIAA-99-3255, 1999.
14. Y. L. Moigne, "Adaptive mesh refinement sensors for Vortex flow simulations," in Proceedings of the European Congress on Computational Methods in Applied Sciences and Engineering (ECCOMAS '04), July 2004.
15. R. Löher and J. D. Baum, "Adaptive h-refinement on 3D unstructured grids for transient problems," International Journal for Numerical Methods in Fluids, vol. 14, no. 12, pp. 1407–1419, 1992.
16. D. Sharov and K. Fujii, "Three-dimensional adaptive bisection of unstructured grids for transient compressible flow computations," AIAA Paper 95-1708, 1995.
17. S. Z. Pirzadeh, "Vortical flow prediction using an adaptive unstructured grid method," inProceedings of the RTO AVT Symposium on Advanced Flow Management: Part A—Vortex Flows and High Angle of Attack for Military Vehicles, Loen, Norway, May 2001.
18. J. Chu and J. M. Luckring, "Experimental surface pressure data obtained on 65° delta wing across reynolds number and mach number ranges," NASA Technical Memorandum 4645, 1996.
19. R. Konrath, A. Schröder, and J. Kompenhans, "Analysis of PIV results obtained for the VFE-265° delta wing configuration at sub- and transonic speeds," in Proceedings of the 24th Applied Aerodynamics Conference, vol. 2, pp. 746–760, San Francisco, Calif, USA, June 2006.

Chapter 3

COMPUTATIONAL FLUID DYNAMICS MODELING TO IMPROVE NATURAL FLOW RATE AND SWEET PEPPER PRODUCTIVITY IN GREENHOUSE

W. Limtrakarn,[1] P. Boonmongkol,[1] A. Chompupoung,[2] K. Rungprateepthaworn,[3] J. Kruenate,[4] and P. Dechaumphai[5]

[1]Department of Mechanical Engineering, Thammasat University, Pathumthani 12120, Thailand

[2]Royal Project Foundation, Chiang Mai 50200, Thailand

[3]National Electronics and Computer Technology Center, Klong Luang, Pathumthani 12120, Thailand

[4]National Metal and Materials Technology Center, Klong Luang, Pathumthani 12120, Thailand

[5]Department of Mechanical Engineering, Chulalongkorn University, Bangkok 10330, Thailand

ABSTRACT

Natural flow rate and sweet peppers productivity in tropical greenhouse are improved by CFD simulation is the main objective of this research work. Most of the greenhouse types today are in the arch shape. To develop an improved greenhouse structure for the region, the arch type was built and used as the control model. Mae Sar Mai agriculture research station under the royal project foundation was selected as the field test site. Temperature sensors with data logger were installed to monitor variation of temperature inside the greenhouse. The measured temperature data were used as the boundary conditions for the CFD analysis. A new greenhouse model with two-step roof shape was designed and the air flow behavior was simulated by using CFD. Regarding CFD results the air flow rate of the new model is about 39% higher than that of old model. The maximum temperature of the new model is lower than that of the old one. The sweet paper growths in both greenhouse models were measured and compared. Results show that the new model obtains 4°C lower maximum temperature in day time, 97% in number and 90% in weight higher the first grade pepper productivity than the old one.

INTRODUCTION

During the past decades, several types of greenhouses with different shapes have been used for crop growing in Thailand. These are the Ridge and Furrow, Saw-tooth, Venlo, Quonset, Gothic arch, and Tunnel shapes. Most of them were originally designed and used in the European countries. These models are introduced and constructed in the northern region of Thailand. In the past, most greenhouses cannot yield good production as they do in the cool climate countries. Structure configurations have been modified so that they are suitable for the high-land area. Plant production and growth in tropical countries strongly depend on the greenhouse shapes and their dimensions. They also depend on the temperature, humidity, ventilation, solar heat transmission, covering materials, CO_2, and so forth [1, 2]. Many research works have been performed to improve their performance by studying natural ventilation inside [3–11]. Few of them studied the temperature, pressure and flow patterns by using the CFD analysis [12–14] and effect from the insect-proof screen installation [7, 15].

Plastic greenhouse is mostly used in the northern region of Thailand because of its low cost, ease of installation, and maintenance. There are some research works that study the plant production in plastic greenhouses [2, 16, 17]. Recently, some works have investigated the micro-climate effect for increasing plant productivity in the greenhouses [18, 19].

The main objective of this study is to redesign the roof shape of a greenhouse for reducing overall inside temperature level. The study employs the CFD analysis simulation to help increasing natural flow rate and sweet pepper productivity. Greenhouse structure is made of steel, covered with plastic sheet, and built at Mae Sa Mai village (900 meters above the sea level) of the royal project, Chiang Mai province in Thailand. Sweet pepper crop is a prospect plant, which grows from April to August. In this paper, the arch-type greenhouse is selected as a controlled model. Ten temperature and relative humidity (T&H) sensors are installed to measure the climate change inside the greenhouse. These measured data are used as the boundary conditions for predicting the internal air flow behavior. To arrive at a design shape of the greenhouse that can provide lower interior temperature, the CFD analysis simulation is used. The new model is designed with two-step roof configuration. Prototypes of the old and new models are constructed and with temperature and humidity sensors installed inside. Sweet pepper nurseries are then moved into these two models. Interior temperature and sweet pepper production of two models are examined and recorded.

MATERIALS AND METHODS

Greenhouse Configuration and Measurements

Survey of existing greenhouses at Mae Sa Mai village in Chiang Mai province shows that there are four types of greenhouses used. About 80% of them are the arch type, 16% are the Saw-tooth type, 2% are the Quonset type, and 2% are the Ridge and Furrow type. Thus, the arc type greenhouse was selected in this research work. Figure 1 shows the arch-type configuration, which was built at the latitude 18° 52′ North, longitude 98° 51′ East, and 900 meters above the sea level. Its dimensions are 6, 5 and 12 m in length, height and width, respectively. The roof is covered by a plastic with UV filter. An insect screen was installed for all sides and the greenhouse door. Air temperature and relative humidity were measured every 5 minutes using the thermistor sensors and data logger DLT-1 10000 M at the roof, the middle, and on the ground. The data collection system sent all data directly to a data server in Bangkok using GSM/General Packet Radio Service (GPRS) gateway connection. Figure 2 shows locations of the nine sensors inside the greenhouse, with one weather station above roof and one sensor outside the greenhouse. The measurement data were collected during January through August.

(a)

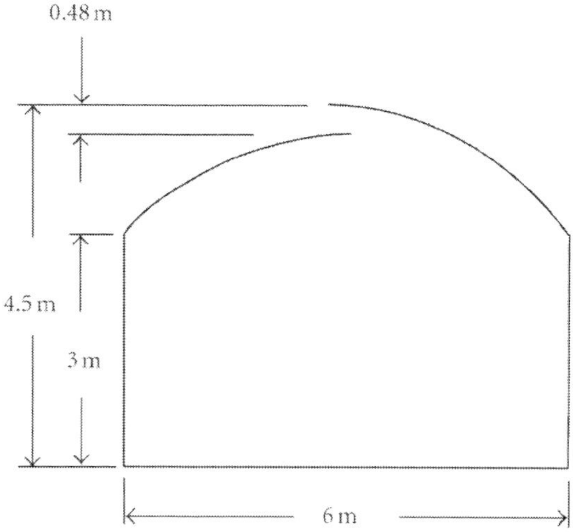

(b)

Figure 1: Arch-type greenhouse configuration.

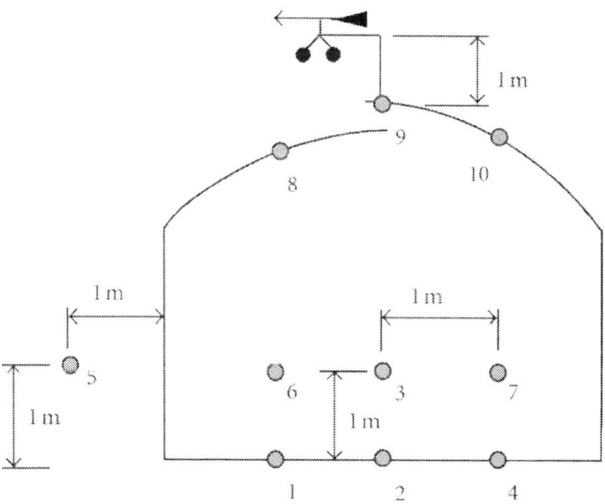

(a) T & H sensors location

(b) Weathering station location

Figure 2: T & H sensors and weathering station location.

CFD Numerical Model

The commercial CFD software, ANSYS/CFX, was used to generate a mesh model for predicting air flow behavior inside. The air flow behavior is governed by the Reynolds-Averaged Navier-Stokes differential equations in the form

$$\frac{\partial \bar{u}}{\partial x} + \frac{\partial \bar{v}}{\partial y} + \frac{\partial \bar{w}}{\partial z} = 0, \tag{1}$$

$$\begin{aligned} &\rho\left(\frac{\partial \bar{u}}{\partial t} + \bar{u}\frac{\partial \bar{u}}{\partial x} + \bar{v}\frac{\partial \bar{u}}{\partial x} + \bar{w}\frac{\partial \bar{u}}{\partial x}\right) \\ &= -\frac{\partial \bar{p}}{\partial x} + \frac{\partial}{\partial x}\left[\mu\left(\frac{\partial \bar{u}}{\partial x} + \frac{\partial \bar{u}}{\partial x}\right) - \rho u'u'\right] \\ &+ \frac{\partial}{\partial y}\left[\mu\left(\frac{\partial \bar{u}}{\partial y} + \frac{\partial \bar{v}}{\partial x}\right) - \rho u'v'\right] \\ &+ \frac{\partial}{\partial z}\left[\mu\left(\frac{\partial \bar{u}}{\partial z} + \frac{\partial \bar{w}}{\partial x}\right) - \rho u'w'\right], \end{aligned} \tag{2}$$

$$\rho\left(\frac{\partial\overline{v}}{\partial t}+\overline{u}\frac{\partial\overline{v}}{\partial x}+\overline{v}\frac{\partial\overline{v}}{\partial x}+\overline{w}\frac{\partial\overline{v}}{\partial x}\right)$$
$$=-\frac{\partial\overline{p}}{\partial y}+\frac{\partial}{\partial x}\left[\mu\left(\frac{\partial\overline{v}}{\partial x}+\frac{\partial\overline{u}}{\partial y}\right)-\rho v'u'\right]$$
$$+\frac{\partial}{\partial y}\left[\mu\left(\frac{\partial\overline{v}}{\partial y}+\frac{\partial\overline{v}}{\partial x}\right)-\rho v'v'\right] \qquad (3)$$
$$+\frac{\partial}{\partial z}\left[\mu\left(\frac{\partial\overline{v}}{\partial z}+\frac{\partial\overline{w}}{\partial x}\right)-\rho v'w'\right],$$

$$\rho\left(\frac{\partial\overline{w}}{\partial t}+\overline{u}\frac{\partial\overline{w}}{\partial x}+\overline{v}\frac{\partial\overline{w}}{\partial x}+\overline{w}\frac{\partial\overline{w}}{\partial x}\right)$$
$$=-\frac{\partial\overline{p}}{\partial z}+\frac{\partial}{\partial x}\left[\mu\left(\frac{\partial\overline{w}}{\partial x}+\frac{\partial\overline{u}}{\partial z}\right)-\rho w'u'\right]$$
$$+\frac{\partial}{\partial y}\left[\mu\left(\frac{\partial\overline{w}}{\partial y}+\frac{\partial\overline{v}}{\partial x}\right)-\rho w'v'\right] \qquad (4)$$
$$+\frac{\partial}{\partial z}\left[\mu\left(\frac{\partial\overline{w}}{\partial z}+\frac{\partial\overline{w}}{\partial x}\right)-\rho w'w'\right],$$

$$\frac{\partial}{\partial t}\left[\rho\left(e+\frac{\overline{V}^2}{2}\right)\right]+\vec{\nabla}\cdot\left[\rho\left(e+\frac{\overline{V}^2}{2}\right)\vec{V}\right]$$
$$=\rho\overline{Q}+\frac{\partial}{\partial x}\left(k\frac{\partial T}{\partial x}\right)+\frac{\partial}{\partial y}\left(k\frac{\partial T}{\partial y}\right)$$
$$-\frac{\partial(up)}{\partial x}-\frac{\partial(vp)}{\partial y}+\frac{\partial(\overline{u}\sigma_x)}{\partial x}+\frac{\partial(\overline{u}\tau_{yx})}{\partial y} \qquad (5)$$
$$+\frac{\partial(\overline{v}\tau_{xy})}{\partial x}+\frac{\partial(\overline{v}\sigma_y)}{\partial y}+\rho\vec{f}\cdot\vec{V},$$

where $\overline{u}$, $\overline{v}$, $\overline{w}$ are the mean velocity components in x, y, and z directions, respectively, while u′, v′, w′ are their fluctuating velocity components. In the above equations, ρ is the fluid density ρ is the fluid pressure, e is the total energy, σ and τ are the normal and shear stress components, respectively.

Equation (1) represents the conservation of mass, while (2)–(4) are the conservation of momentums in *x*, *y*, and *z* directions, respectively. The last equation, (5), is the conservation of energy. These equations were solved by the finite volume method. The standard *k*-ε turbulence model was employed to represent the ventilation phenomena [20, 21]. The instantaneous fluid velocity is sum of the mean and fluctuating components of the turbulent velocity. In the standard κ-ε turbulence model, the turbulent viscosity is computed as a function of the turbulent energy (κ) and dissipation rate of the turbulent kinetic energy (ε).

Computational Procedure

The air properties, as shown in Table 1, are the density at 900 m above the sea level, the specific heat, thermal conductivity coefficient, and viscosity at outside temperature. Boundary conditions were selected from the measure data of temperature sensors. Table 2 shows the average measure data at 2:00 pm on 11 February at the three heights and outside the model. Roof and ground temperatures were used as Dirichlet boundary conditions, while initial temperature was the outside temperature. The external wind speed and direction were measured from the weather station. It is noted that the wind speed is close to zero about 66% during the day time. Natural ventilation condition was selected in the analysis because the airflow outside the greenhouse is near a stagnation condition.

Table 1: Air properties at Mae Sa Mai village.

Air Properties	
Density	1.185 kg/m^3
Specific heat	1004.4 J/kg·k
Thermal conductivity coefficient	0.0261 W/m·k
Viscosity	1.831×10^{-5} kg/m·s

Table 2: Averagemeasured data at different heights on 11 February.

Measured location	Temperature (°C)
Roof	39.6
1 m above ground	26.4
Ground	32
Outside	25

RESULTS AND DISCUSSION

Measurement System and Data

The temperature, relative humidity, and wind speed data were collected at every 5 minutes and sent through the GSM/GPRS system as well as recorded on a computer server in Bangkok. The data bank can be accessed online by the internet system. Figure 3 shows the average greenhouse temperature that varies with time during 24 hours on 11 February. The greenhouse started receiving the solar heating at 7 AM, and the heat was accumulated during the day time. Temperature was above 24°C between 9 AM 5 PM. The maximum temperature of about 31°C occurred between 1 and 4 pm. It is noted that the sweet pepper crop is originated from cold regions where suitable temperature is about 24°C during the day time and 17°C during the night time. This data implies that the greenhouse needs better ventilation to release accumulated heat during the day time. In this paper, the greenhouse configuration was redesigned to provide lower temperature. Figure 4 shows the wind chart that represents wind speed and its direction. Maximum wind speed is at 3 km/hr and its direction is 347 NW. The wind speed is near zero about 66% during the day time.

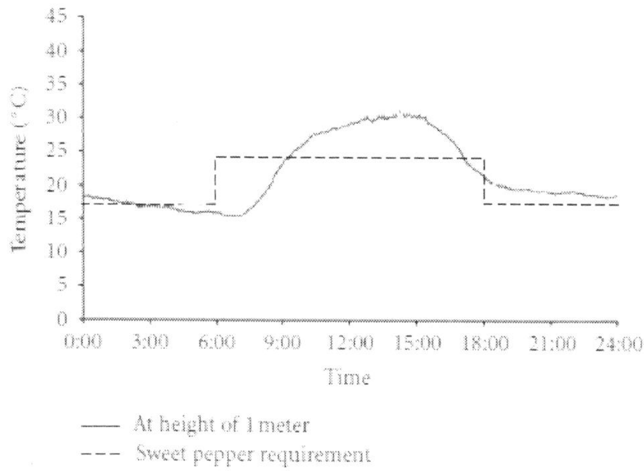

Figure 3: Average temperature of arch-type greenhouse on 11 February.

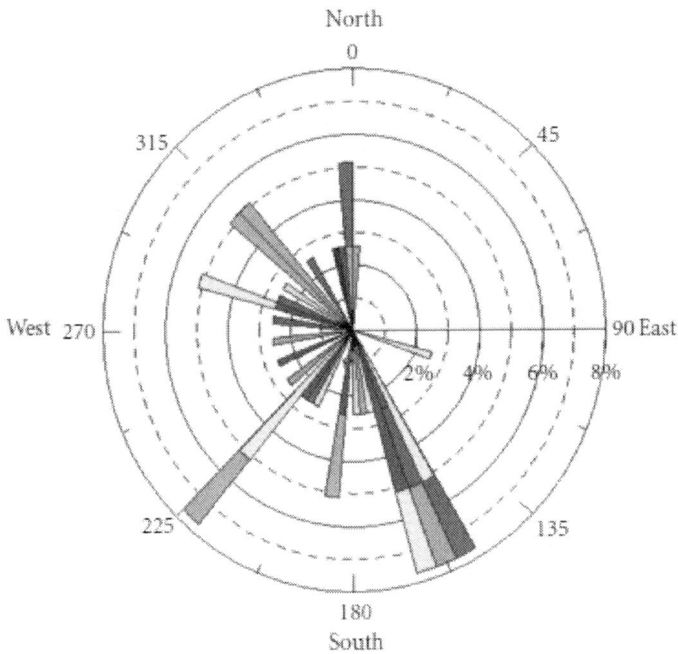

Figure 4: Wind speeds and directions on 11 February.

Simulation Results

With the arc-type greenhouse configuration, the CFD analysis for the air flow behavior was performed. CFD mesh is constructed and consists of 11,000 nodes. Figure 5 shows the air flow velocity vectors inside the greenhouse and near the roof opening. Hot air moves upward and flows out from the greenhouse through the roof opening. Because of small opening at roof, there are large circulation under right roof, medium, and small circulation above floor from left to right. Heat is then accumulated all day and increases the temperature level inside the greenhouse. From the CFD result, it is implied that large circulation under right roof is the major factor in heat accumulation inside greenhouse.

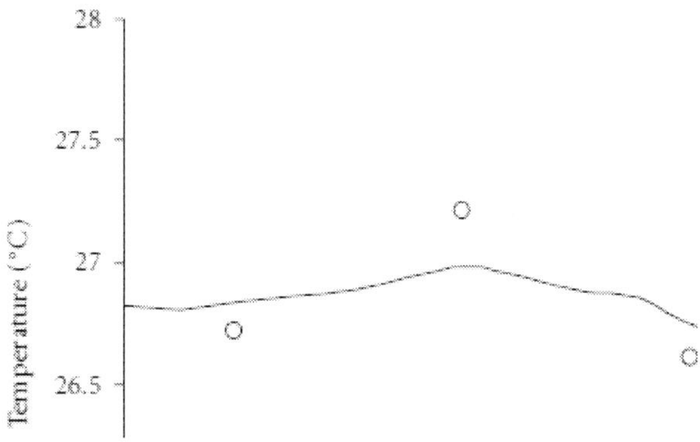

Figure 5: Air flow velocity vectors in the arch-type greenhouse.

Figure 6(a) shows the temperature distribution at 1 meter height above the ground as compared to the measured data. The temperatures at different heights were plotted as shown in Figure 6(b). The temperature results are agreed with the measured data. Regarding high temperature problem in greenhouse, the new design does focus on reduction of accumulated heat using CFD.

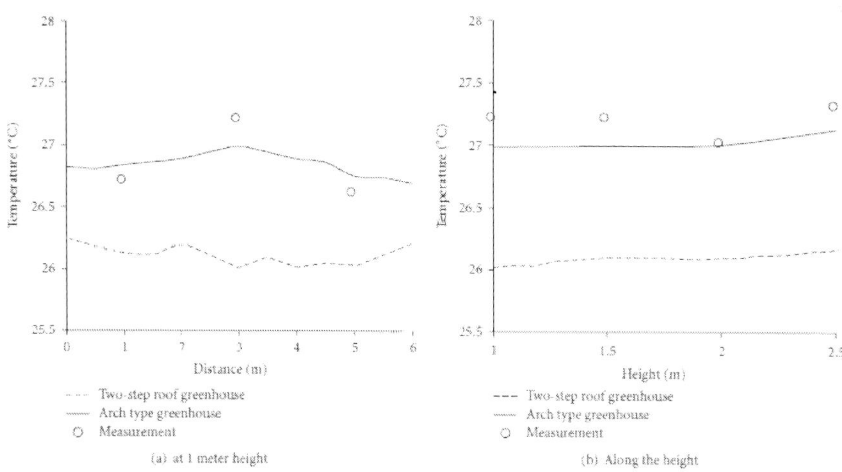

Figure 6: Comparative temperatures between the simulated and measured data.

Based on this concept the roof opening distance is considered. For the redesign and optimized greenhouse configuration, the two-step roof greenhouse shape was selected as shown in Figure 7. Additional roof or second roof is added above the original roof with higher opening distance. Then redesign model is constructed and generated CFD mesh. CFD analysis is then used to solve flow behavior of redesign GH model. Several redesign models were simulated by using CFD analysis to optimize opening distance. The optimized CFD mesh consists of 14,000 nodes. Grid independence is also considered in all CFD model.

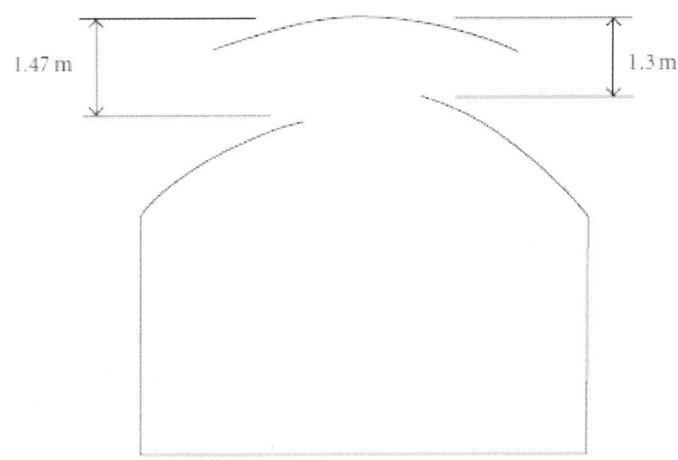

Figure 7: Two-step roof greenhouse configuration.

Figure 8 shows the air flow velocity vectors inside and near the roof openings of the new greenhouse shape. Large circulation under right roof in original model is disappeared in redesign model result. At the roof openings, the air flow rate of the redesign model is about 39% higher than that of old model. Temperature at 1 meter height and along the height of the greenhouse were plotted and compared as shown in Figures 6(a) and 6(b), respectively. At 1 meter height, the temperature in the redesign model is lower than that of the old model. The two-step roof model was then constructed and the measurement sensors were installed in the same locations as used in the old model.

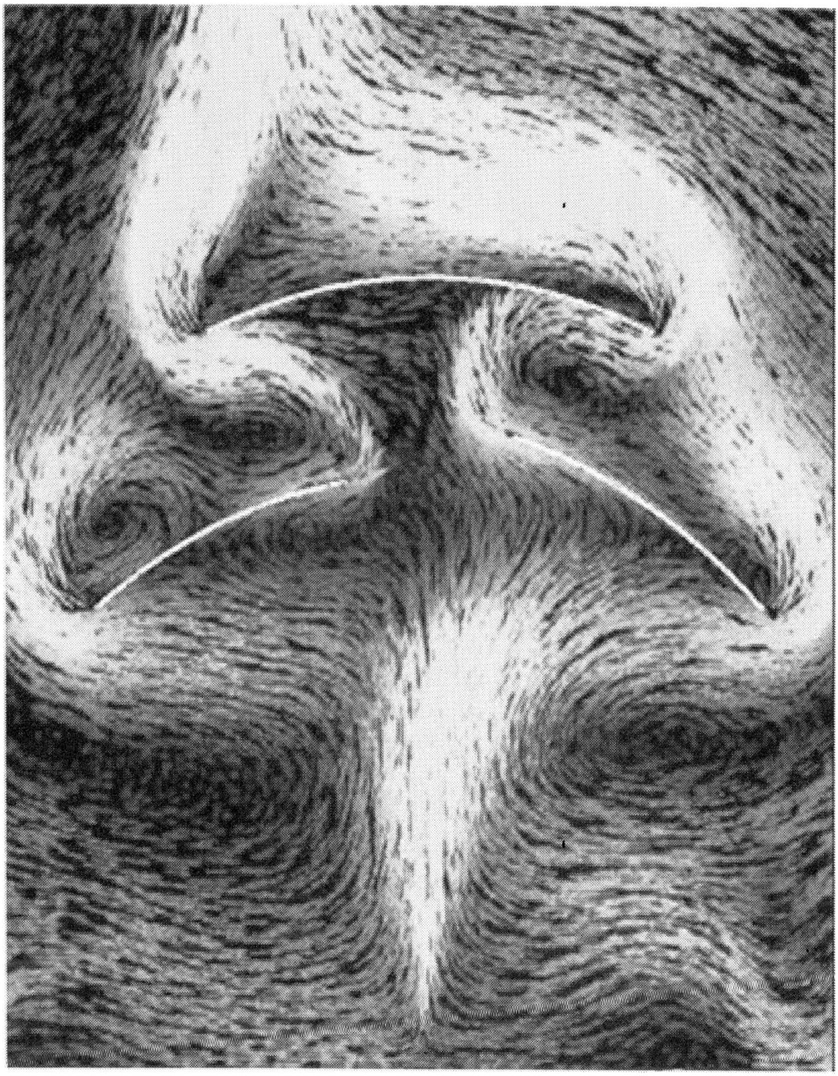

Figure 8: Air flow velocity vectors in the two-step roof greenhouse.

Figure 9 shows the comparison of measured temperatures between two models. The redesign model yields a lower temperature than the old model about 4°C during the day time and higher about 1°C during the night time.

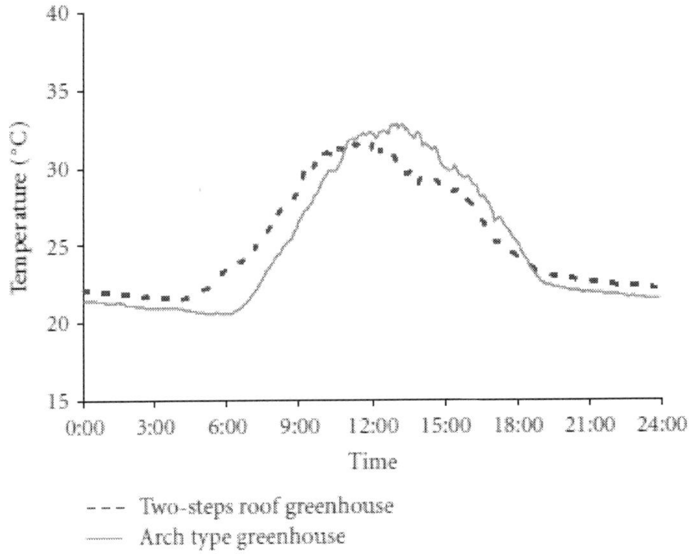

Figure 9: Comparative measured temperatures of the two greenhouse models.

Productivity Results

Sweet pepper crop was plant between May and August. In such period, the plant growth and its productivity were collected 5 times as shown in Table 3. Sweet pepper crop is classified in 2 grades. The first-grade is for those with high quality, while the second grade is with fair quality. The first grade productivities obtained from the new two-step roof, and the old models are 97% and 90%, respectively. The second-grade productivities obtained from the two models are 64% and 70%, respectively.

Table 3: Plant growth and productivity of sweet pepper crop.

Item	d/m/y	First grade				Second grade			
		Two-step roof GH		Arch-type GH		Two-step roof GH		Arch-type GH	
		Number of fruit	Total Weight	Number of fruit	Total Weight	Number of fruit	Total Weight	Number of fruit	Total Weight
1	26/6/08	76	15.2	40	8	10	2.2	7	1.6
2	3/7/08	54	10.8	37	7.4	63	12.6	42	8.4
3	10/7/08	85	17	38	7.5	65	13	40	8.1
4	17/7/08	18	3	8	2	49	9	20	3
5	27/7/08	35	7	15	3	12	2	13	2
	Mean	48	9.45	24	4.975	47	9.15	29	5.375

Concluding Remarks

The CFD simulation is used to enhance natural flow rate and sweet pepper productivity in tropical greenhouses. The CFD technique helped redesigning greenhouse configuration to achieve better ventilation reduces heat storage and temperature inside. The existing and popular greenhouse type, that is, the arch type, at Mae Sa Mai village was selected as the control greenhouse model. The measurement system was installed and the data were transmitted and recorded in a remote server by GSM/GPRS system. A new design configuration with two-step roof was selected. The maximum temperature in new design model is less than that of the old model. At the roof opening section, the air flow rate of new model is 39% higher than that of old model. The two-step roof greenhouse was constructed and a measurement system was installed. Sweet pepper crop was planted inside the old and new models. The maximum temperature measured in the new model is about 4°C less than that of the old model during the day time. The simulated CFD and the measured temperatures are in good agreement. The new design model produced the first-grade sweet pepper productivity about 97% in number and 90% in weight higher than those from the old model. The results have demonstrated the applicability and advantages of the CFD simulation for redesigning the greenhouse configuration to enhance natural flow rate and sweet pepper productivity.

Acknowledgments

The authors are pleased to acknowledge the Thailand Commission on Higher Education of Thailand (the National Research University Project), the National Science and Technology (NRCT), the Thailand Research Fund (TRF), Royal Project Foundation, Thammasat University, and the National Metal and Materials Technology Center (MTEC) for supporting this paper.

REFERENCES

1. B. Von Elsner, D. Briassoulis, D. Waaijenberg et al., "Review of structural and functional characteristics of greenhouses in European Union countries, part II: typical designs," Journal of Agricultural

Engineering Research, vol. 75, no. 2, pp. 111-126, 2000.

2. B. Cemek, Y. Demir, S. Uzun, and V. Ceyhan, "The effects of different greenhouse covering materials on energy requirement, growth and yield of aubergine," Energy, vol. 31, no. 12, pp. 1444-1452, 2006.
3. T. Boulard, G. Papadakis, C. Kittas, andM. Mermier, "Air flow and associated sensible heat exchanges in a naturally ventilated greenhouse," Agricultural and Forest Meteorology, vol. 88, no. 1-4, pp. 111-119, 1997.
4. Mistriotis, C. Arcidiacono, P. Picuno, G. P. A. Bot, and G. Scarascia-Mugnozza, "Computational analysis of ventilation in greenhouses at zero- and low-wind-speeds," Agricultural and Forest Meteorology, vol. 88, no. 1-4, pp. 121-135, 1997.
5. M. Teitel and J. Tanny, "Natural ventilation of greenhouses: experiments and model," Agricultural and Forest Meteorology, vol. 96, no. 1-3, pp. 59-70, 1999.
6. S. Wang, T. Boulard, and R. Haxaire, "Air speed profiles in a naturally ventilated greenhouse with a tomato crop," Agricultural and Forest Meteorology, vol. 96, no. 4, pp. 181-188, 1999.
7. O. F. Nielsen, "Natural ventilation of a greenhouse with top screen," Biosystems Engineering, vol. 81, no. 4, pp. 443-451, 2002.
8. M. Kacira, S. Sase, and L. Okushima, "Effects of side vents and span numbers on wind-induced natural ventilation of a gothic multi-span greenhouse," Japan Agricultural Research Quarterly, vol. 38, no. 4, pp. 227-233, 2004.
9. J. P. Parra, E. Baeza, J. I. Montero, and B. J. Bailey, "Natural ventilation of parral greenhouses," Biosystems Engineering, vol. 87, no. 3, pp. 355-366, 2004.
10. F. J. Jim´enez-Hornero, E. Guti´errez De Rav´e, R. Hidalgo, and J. V. Gir´aldez, "Numerical study of the natural airflow in greenhouses using a two-dimensional lattice model," Biosystems Engineering, vol. 91, no. 2, pp. 219-228, 2005.
11. Kumar and G. N. Tiwari, "Thermal modeling of a natural convection greenhouse drying system for jaggery: an experimental validation," Solar Energy, vol. 80, no. 9, pp. 1135-1144, 2006.
12. K. Kim, G. A. Giacomelii, J. Y. Yoon et al., "CFD modeling to improve the design of a fog system for cooling greenhouses," Japan Agricultural Research Quarterly, vol. 41, no. 4, pp. 283-290, 2007.
13. S. A. O. Khaoua, P. E. Bournet, C. Migeon, T. Boulard, and G. Chass´eriaux, "Analysis of greenhouse ventilation efficiency based on computational fluid dynamics," Biosystems Engineering, vol. 95, no. 1, pp. 83-98, 2006.

14. H. Fatnassi, T. Boulard, C. Poncet, and M. Chave, "Optimisation of greenhouse insect screening with computational fluid dynamics," Biosystems Engineering, vol. 93, no. 3, pp. 301-312, 2006.
15. T. Bartzanas, T. Boulard, and C. Kittas, "Numerical simulation of the airflow and temperature distribution in a tunnel greenhouse equipped with insect-proof screen in the openings," Computers and Electronics in Agriculture, vol. 34, no. 1-3, pp. 207-221, 2002.
16. T. Boulard and S.Wang, "Experimental and numerical studies on the heterogeneity of crop transpiration in a plastic tunnel," Computers and Electronics in Agriculture, vol. 34, no. 1-3, pp. 173-190, 2002.
17. Ghoname and M. R. Shafeek, "Growth and productivity of sweet pepper (Capsicum annum L.) grown in plastic house as affected by organic, mineral and bio-N-fertilisers," Journal of Agronomy, vol. 4, no. 4, pp. 369-372, 2005.
18. H. E. S. Fath and K. Abdelrahman, "Micro-climatic environmental conditions inside a greenhouse with a built-in solar distillation system," Desalination, vol. 171, no. 3, pp. 267-287, 2005.
19. C. Kittas and T. Bartzanas, "Greenhouse microclimate and dehumidification effectiveness under different ventilator configurations," Building and Environment, vol. 42, no. 10, pp. 3774-3784, 2007.
20. B. E. Launder and D. B. Spalding, "The numerical computation of turbulent flows," Computer Methods in Applied Mechanics and Engineering, vol. 3, no. 2, pp. 269-289, 1974.
21. M. Kacira, T. H. Short, and R. R. Stowell, "A CFD evaluation of naturally ventilated, multi-span, sawtooth greenhouses," Transactions of the American Society of Agricultural Engineers, vol. 41, no. 3, pp. 833-836, 1998.

Chapter 3

APPLICATION OF COMPUTATIONAL FLUID DYNAMICS TO PRACTICAL DESIGN AND PERFORMANCE ANALYSIS OF TURBOMACHINERY

Vadim Seleznev

[1] Physical and Technical CenterSarov, Nizhny Novgorod Region,, Russia

INTRODUCTION

At the present level of development of long, branched gas transmission networks (GTN), solving the problems of improving safety, efficiency and environmental soundness of operation of industrial pipeline systems calls for the application of methods of numerical simulation. The development of automated devices for technical inspection and process control, and availability of high-performance computer hardware have created a solid technical basis to introduce numerical simulation methods into the industrial practice of GTN analysis and operation. One of the promising approaches for numerical analysis of GTN operating is the development and application of high-accuracy computational fluid dynamics (CFD) simulators of modes of gas mixture transmission through long, branched pipeline systems (CFD-simulator) (Seleznev, 2007).

Actually, a CFD-simulator is a special-purpose software simulating, in "online" and "real time" modes with a high similarity and in sufficient detail, the physical processes of gas mixture transmission through a particular GTN. The development of a CFD-simulator focuses much attention to correctness of simulation of gas flows in the pipelines and to the impact produced by operation of relevant GTN gas pumping equipment (including gas compressor unit (GCU), valves, gas pressure reducers, etc.) and the environment upon the physical processes under study.

From the standpoint of mathematical physics, a CFD-simulator performs numerical simulation of steady and transient, non-isothermal processes of a gas mixture flow in long, branched, multi-line, multi-section gas pipeline network. Such simulation is aimed at obtaining high-accuracy estimates of the actual distribution (over time and space) of fluid dynamics parameters for the full range of modes of gas mixture transmission through the specific GTN in normal and emergency conditions of its operation, as well as of the actual (temporal) distribution of main parameters of GTN equipment operation, which can be expressed as functional dependencies on the specified controls on the GTN and corresponding boundary conditions. Theoretically, the high-accuracy of estimates of gas flow parameters is achieved here due to (Seleznev et al., 2005): (1) minimization of the number and depth of accepted simplifications and assumptions in the mathematical modeling of gas flows through long, branched, multi-section pipelines and gas compressor stations (CS) on the basis of adaptation of complete basic fluid dynamics models, (2) minimization of the number and depth of accepted simplifications and assumptions in the construction of a computational model of the simulated GTN, (3) improving methods for numerical analysis of the constructed mathematical models based upon results of theoretical investigation of their convergence and evaluation of possible errors of solution, (4) taking into account the mutual influence of GTN components in the simulation of its operation, (5) detailed analysis and mathematically formal description of the technologies and supervisor procedures for management of gas mixture transport at the simulated GTN, (6) automated mathematic filtration of occasional and systemic errors in input data, etc.

Input information required for work of a CFD-simulator is delivered from the Supervisory Control and Data Acquisition System

(SCADA-system) operated at the simulated GTN.

CFD-simulator's operating results are used for on-line control of the specific GTN, as well as in short-term and long-term forecasts of optimal and safe modes of gas mixture transport subject to fulfillment of contractual obligations. Also, a CFD-simulator is often used as base software for a hardware and software system for prevention or early detection of GTN failures.

For better illustration of the material presented in this chapter, but without loss of generality, further description of a CFD-simulator will be based on a sample pipeline network of a gas transmission enterprise. For the purpose of modeling, natural gas is deemed to be a homogenous gas mixture. A CFD-simulator of a gas transmission enterprise's GTN is created by combining CS mathematical models into a single model of the enterprise's pipeline system, by applying models of multi-line gas pipelines segments (GPS) (Seleznev et al., 2007). At that, in accordance with their process flow charts, the CS models are created by combining of GCU, dust catcher (DC) and air cooling device (ACD) models by applying mathematical models of connecting gas pipelines (CGP).

In a CFD-simulator, the control of simulated natural gas transmission through the GTN is provided by the following control commands: alteration of shaft rotation frequency of centrifugal superchargers (CFS) of GCU or their startup/shutdown; opening or closing of valves at a CS and valve platforms of multi-line GPS; alteration of the rates of gas consumption by industrial enterprises and public facilities; alteration of the gas reduction program at reduction units; alteration of the operation program at gas distributing stations; change in the program of ACD operating modes, etc. Therefore, simulated control in a CFD-simulator adequately reflects the actual control of natural gas transmission through pipeline networks of the gas transmission enterprise.

Generally, a CFD-simulator can be divided into three interrelated components (elements) (Seleznev et al., 2007). Each of these components is an integral part of the CFS-simulator.

The first system element is a computational scheme of a gas transmission enterprise pipeline system built on the basis of typical segments representing minimum distinctions from a comprehensive topology of an actual system considering the arrangement of valves, the system architecture, laying conditions, the process flow scheme of the system's CS, etc. The second component is a database containing input and operative (current) data on time-dependent (owing to valves operation) system topology, pipeline parameters, process modes and natural gas transmission control principles for an actual gas transmission enterprise. The third component of a CFD-simulator is a mathematical software which operates the first two CFD-simulator elements.

The mathematical software includes (in addition to the computation core) a user interface environment imitating the operation of actual control panels located at gas transmission enterprises control centers in a visual form familiar to operators. This provides for faster training and, for the operator, easier adaptation to the CFD-simulator.

SIMULATION OF MULTI-LINE GPS BY CFD-SIMULATOR

Multi-line GPS are long, branched, multi-section pipelines. For numerical evaluation of parameters of steady and transient, non-isothermal processes of the gas mixture flow in multi-line GPS, a CFD-simulator uses a model developed by tailoring the full set of integral fluid dynamics equations to conditions of the gas flow through long branched pipeline systems. Transform of the 3D integral problem to an equivalent one-dimensional differential problem is implemented by accepting the minimum of required simplifications and projecting the initial system of equations onto the pipeline's geometrical axis. At that, a special attention is given to the adequacy of simulation of pipeline junction nodes where the 3D nature of the gas flow is strongly displayed.

In order to improve the accuracy of the simulations, it is reasonable to use the CFD-simulator in the "online" and "real time" modes for the numerical analysis of the given processes. There are two mathematical models of fluid flow through branched pipeline: heat-conductive model of pipeline junction and nonconductive model of pipeline junction. These models were developed by S. Pryalov and V. Seleznev at the turn of the century. These alternatives differ in a way of simulation of gas heat transfer within pipeline junction. The principle underlying the simulations is to observe the major conservation laws as strictly as possible. In practice the simultaneous implementation of the models makes it possible to find an accurate solution in short time.

The basis for the mathematical models of fluid flow through branched pipeline was the geometrical model of a junction (fig. 1) proposed by S. Pryalov (Seleznev et al., 2005). In this model, volume ${}^{(0)}V$ can be depicted as a right prism with base area S_{base} and height H (see fig. 1a). For the prism lateral surface with linear dimensions ${}^{(n)}$, true is the following relation: ${}^{(n)} = {}^{(n)}f/H$, where $f(n)$ is the cross-sectional area of the pipe adjacent to the junction core ${}^{(0)}V$. It shc d that the summarized volume of the joint is equal to $V = \bigcup_{n=0}^{N} {}^{(n)}V$, where ${}^{(n)}Vn=1...N$ is the volume of an infinitely small section of the pipe adjacent to the junction core $V(0)$ (see fig. 1b). The prism base area can be represented as follows: $S_{base} = {}_{base}\,({}^{1)}$ 2, where ${}_{base}$ is the factor depending on the prism base geometry $^{(0)}V = HS_{base} = = H\varsigma_{base}\left({}^{(1)}f/H\right)^2 = \varsigma_{base}{}^{(1)}f^2/H$lowing formula: , which means that $\lim_{H\to\infty} {}^{(0)}V = \lim_{H\to\infty}\left[\varsigma_{base}{}^{(1)}f^2/H\right] = 0$. The application of this geometrical model enabled us to approximate compliance with mass, momentum and energy conservation laws at the pipelines junction.

Simplifications and assumptions used to construct the heat-conductive model of pipeline junction include the following: (1) when gas mixture flows join together, pressure can change with time, but at each time step it will have the same value at the boundaries of the pipeline junction, (2) the simulations take account of «downwind" heat and mass exchange due to heat conduction and diffusion, (3) in the pipeline junction, the gas mixture instantaneously becomes ideally uniform all over the pipeline junction volume ${}^{(0)}V$ (see fig.

1b), (4) effects of gas mixture viscosity in the pipeline junction (inside the volume ${}^{(0)}V$) can be ignored, (5) there are no heat sources in ${}^{(0)}V$ (inside the volume ${}^{(0)}V$), (6) pipeline diameters near the pipeline junction are constant.

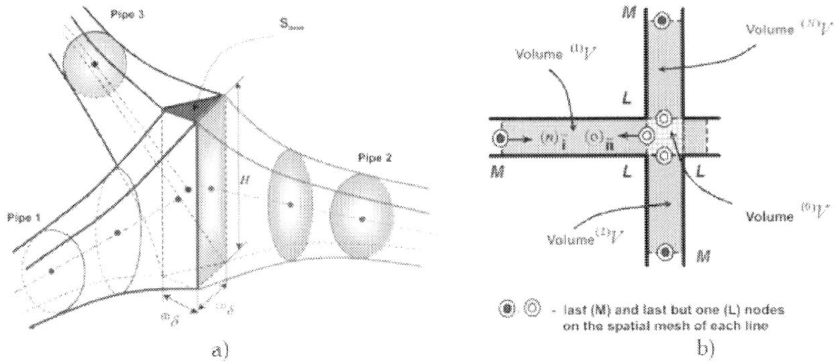

Figure 1. A schematic of a pipeline junction (a – 3D drawing; b – diagram)

Then, the heat-conductive fluid dynamics model of a transient, non-isothermal, turbulent flow of a viscous, chemically inert, compressible, heat-conductive, multi-component gas mixture through multi-line GPS which consist of pipes of round cross-sections and rigid rough heat-conductive walls is represented in the following way (Seleznev et al., 2005):

for each pipe adjacent to the junction node

$$\frac{\partial(\rho f)}{\partial t} + \frac{\partial}{\partial x}(\rho w f) = 0$$

$$\frac{\partial}{\partial t}(\rho Y_m f) + \frac{\partial}{\partial x}(\rho Y_m w f) - \frac{\partial}{\partial x}\left(\rho f D_m \frac{\partial Y_m}{\partial x}\right) = 0 m = \overline{1 N_S - 1} Y_{N_S} = 1 - \sum_{m=1}^{N_S-1} Y_m$$

$$\frac{\partial(\rho w f)}{\partial t} + \frac{\partial(\rho w^2 f)}{\partial x} = -f\left(\frac{\partial p}{\partial x} + g\rho \frac{\partial z_1}{\partial x}\right) - \frac{\pi}{4}\lambda \rho w |w| R$$

$$\frac{\partial}{\partial t}\left[\rho f\left(\varepsilon+\frac{w^2}{2}\right)\right]+\frac{\partial}{\partial x}\left[\rho w f\left(\varepsilon+\frac{w^2}{2}\right)\right]=-\frac{\partial}{\partial x}(p w f)-\rho w f g\frac{\partial z_1}{\partial x}-$$
$$-p\frac{\partial f}{\partial t}+Qf+\frac{\partial}{\partial x}\left(kf\frac{\partial T}{\partial x}\right)-\Phi(TT_{\mathrm{am}})+\frac{\partial}{\partial x}\left(\rho f\sum_{m=1}^{N_S}\varepsilon_m D_m\frac{\partial Y_m}{\partial x}\right)$$

for each of the junction nodes

$$\frac{\partial\rho}{\partial t}+\sum_{n=1}^{N}{}^{(n)}\Theta^{(n)}\left(\frac{\partial(\rho w)}{\partial x}\right)=0$$

$$\frac{\partial(\rho Y_m)}{\partial t}+\sum_{n=1}^{N}{}^{(n)}\left(\frac{\partial(\rho Y_m w)}{\partial x}\right){}^{(n)}\Theta-\sum_{n=1}^{N}{}^{(n)}\left[\frac{\partial}{\partial x}\left(\rho D_m\frac{\partial Y_m}{\partial x}\right)\right]{}^{(n)}\Theta=0$$
$$m=\overline{1N_S-1}\;{}^{(n)}Y_{N_S}=1-\sum_{m=1}^{N_S-1}{}^{(n)}Y_m$$

$${}^{(n)}\left(\frac{\partial(\rho w)}{\partial t}\right)+{}^{(n)}\left(\frac{\partial(\rho w^2)}{\partial x}\right)=-{}^{(n)}\left(\frac{\partial p}{\partial x}\right)-g\rho^{(n)}\left(\frac{\partial z_1}{\partial x}\right)-0.25^{(n)}\left(\frac{\lambda\rho w|w|}{R}\right)\;n=\overline{1N}$$

$$\frac{\partial(\rho\varepsilon)}{\partial t}+\sum_{n=1}^{N}{}^{(n)}\left(\frac{\partial(\rho\varepsilon w)}{\partial x}\Theta\right)=-\sum_{n=1}^{N}{}^{(n)}\left(p\frac{\partial w}{\partial x}\Theta\right)+0.25\sum_{n=1}^{N}{}^{(n)}\left(\frac{\lambda\rho|w|^3\Theta}{R}\right)+$$
$$+Q+\sum_{n=1}^{N}{}^{(n)}\left(\left[\frac{\partial}{\partial x}\left(k\frac{\partial T}{\partial x}\right)\right]\Theta\right)-\sum_{n=1}^{N}{}^{(n)}\left(\frac{\Phi(TT_{\mathrm{am}})}{f}\Theta\right)+\sum_{n=1}^{N}\sum_{m=1}^{N_S}{}^{(n)}\left(\left[\frac{\partial}{\partial x}\left(\rho\varepsilon_m D_m\frac{\partial Y_m}{\partial x}\right)\right]\Theta\right)$$

$${}^{(n)}T={}^{(\xi)}T\,\varepsilon={}^{(n)}\varepsilon={}^{(\xi)}\varepsilon^{(n)}(\varepsilon_m)={}^{(\xi)}(\varepsilon_m)\,\rho={}^{(n)}\rho={}^{(\xi)}\rho^{(n)}p={}^{(\xi)}p$$
$${}^{(n)}k={}^{(\xi)}k^{(n)}(D_m)={}^{(\xi)}(D_m)\,Y_m={}^{(n)}(Y_m)={}^{(\xi)}(Y_m)^{(n)}(z_1)={}^{(\xi)}(z_1)$$
for any $n\xi\in\overline{1N}$ and $m\in\overline{1N_S}$

$$\sum_{n=1}^{N} {}^{(n)}(wfs) = 0 \sum_{n=1}^{N} {}^{(n)}\left(\frac{\partial T}{\partial x} fs\right) = 0 \sum_{n=1}^{N} {}^{(n)}\left(\frac{\partial Y_m}{\partial x} fs\right) = 0$$

$$^{(n)}s = \begin{cases} 1 if \left({}^{(0)}\vec{\mathbf{n}} \cdot {}^{(n)}\vec{\mathbf{i}}\right) 0 \\ -1 if \left({}^{(0)}\vec{\mathbf{n}} \cdot {}^{(n)}\vec{\mathbf{i}}\right) 0 \end{cases} \quad {}^{(n)}\Theta = \frac{{}^{(n)}V}{V} = \frac{{}^{(n)}f_L {}^{(n)}\gamma}{\sum_{k=1}^{N}\left({}^{(k)}f_L {}^{(n)}\gamma\right)} 0^{(n)}\Theta 1 \sum_{n=1}^{N} {}^{(n)}\Theta = 1$$

equation of state (EOS) and additional correlations:

$$p = p(\{S_{\text{mix}}\})\varepsilon = \varepsilon(\{S_{\text{mix}}\})k = k(\{S_{\text{mix}}\})\varepsilon_m = \varepsilon_m(\{S_{\text{mix}}\})D_m = D_m(\{S_{\text{mix}}\})m = 1N_S$$
$$T_1 = T_2 = \ldots = T_{N_S} = T$$

where is the density of the gas mixture; f is the flow cross-sectional area of pipeline; t is time (marching variable); x is the spatial coordinate over the pipeline's geometrical axis (spatial variable); w is the projection of the pipeline flow cross-section averaged vector of the mixture velocity on the pipeline's geometrical axis (on the assumption of the developed turbulence); Y_m is a relative mass concentration of the m component of the gas mixture); D_m is a binary diffusivity of component m in the residual mixture; N_S is the number of omponents of the homogeneous gas mixture; p is the pressure in the gas mixture; g is a gravitational acceleration modulus; z_1 is the coordinate of the point on the pipeline's axis, measured, relative to an arbitrary horizontal plane, upright; ϖ is the Pythagorean number; is the friction coefficient in the Darcy – Weisbach formula; $R -\sqrt{f/\varpi}$ is the pipe›s internal radius; is specific (per unit mass) internal energy of the gas mixture; Q is specific (per unit volume) heat generation rate of sources; k is thermal conductivity; T is the temperature of gas mixture; m is specific (per unit mass) internal energy of the m component; T_m is the temperature of the m component; N is the number of pipes comprising one junction (see (5–11)); $s(n)$ (n) are auxiliary functions (re ${}^{(0)}\vec{\mathbf{n}}$ ${}^{(n)}\vec{\mathbf{i}}$ see fig. 1b below); $\{S_{\text{mix}}\}$is a set of parameters of gas mixture. Function (TT_{am}) is defined by the law of heat transfer from the pipe to the environment and expresses the aggregate heat flow through the pipe walls along perimeter of the flow cross-section with area f ($(TT_{am})0$ is cooling), T_{am} is the ambient temperature. To denote the relationship of a value to the pipe numbered by n, we use a parenthesized superscript on the left side of the value, e.g.: (n). In equations (1–12), we use physical magnitudes averaged across the pipeline's flow cross-section. The set of equations (1–12) is supplemented by the boundary conditions and conjugation conditions. As conjugation conditions it is possible to specify boundary conditions simulating a complete rupture of the pipeline and/or its shutoff, operation of valves, etc.

As was stated above, the energy equations (4) and (8) comprise function (TT_{am}) describing the heat exchange between the environment and natural gas in the course of its pipeline transmission. The space-time distribution of function (TT_{am}) is defined, in the CFD-simulator, at specified time steps of the numerical analysis of parameters of the transient mode of gas transmission by solving a series of conjugate 2D or 3D problems of heat exchange between the gas flow core and the environment (Seleznev et al., 2007).

Simulation of steady processes of gas mixture flow through multi-line GPS is a less complicated task compared to (1–12). These models can be easily derived by simplifying the set of equations (1–12).

When simulating the gas mixture flow through real multi-line GPSs, one uses meshes with a spatial cell size of 10m to 10,000m. As a result, a smooth growth (or decrease) of temperature in a span of about $10{-}5m$ will be simulated as a temperature jump. Difference "upwind" equations make it possible to find a solution, which is quite adequate for the process at issue, almost without any impact on the convergence and accuracy of the resulting solution. On the contrary, the schemes that use the principles of central differences as applied to this process can yield solutions with difference oscillations. This may impair the accuracy of such simulations.

To overcome this drawback, S. Pryalov and V. Seleznev in 2008 suggested using the nonconductive model of pipeline junction. Downwind (and upwind) heat conduction and diffusion in this model are ignored. There will be a temperature discontinuity on gas mixture transition through the pipeline junction node, but this dependence will be monotone along each pipeline.

The list of additional simplifications in setting up the nonconductive model of pipeline junction includes only one item:

the model ignores the downwind heat and mass exchange in the gas mixture due to heat conduction and diffusion. The temperature of the gas mixture at the outlet boundaries of inlet pipelines is defined only by the mixture flow parameters (mainly, by the temperature) inside these pipelines.

As there is no downwind heat transfer mechanism from the volume ${}^{(0)}V$ to the inlet pipelines, the temperature of the gas mixture at these boundaries may differ from the temperature inside the volume ${}^{(0)}V$. On the other hand, the outlet boundaries of the inlet pipelines are also boundaries of the volume ${}^{(0)}V$. For this reason, it does not seem to be correct to say that the mixture is uniformly intermixed all over the volume.

Thus, the nonconductive model of pipeline junction of a transient, non-isothermal, turbulent flow of a viscous, chemically inert, compressible, multi-component gas mixture through multi-line GPS which consist of pipes of round cross-sections and rigid rough heat-conductive walls contains equations (1), (3), (7), (11) and the following equations:

for each pipe adjacent to the junction node

$$\frac{\partial}{\partial t}(\rho Y_m f)+\frac{\partial}{\partial x}(\rho Y_m wf)=0m=\overline{1N_S-1}Y_{N_S}=1-\sum_{m=1}^{N_S-1}Y_m$$

$$\frac{\partial}{\partial t}\left[\rho f\left(\varepsilon+\frac{w^2}{2}\right)\right]+\frac{\partial}{\partial x}\left[\rho wf\left(\varepsilon+\frac{w^2}{2}\right)\right]=$$
$$=-\frac{\partial}{\partial x}(pwf)-\rho wfg\frac{\partial z_1}{\partial x}-p\frac{\partial f}{\partial t}+Qf-\Phi(TT_{\mathrm{am}})$$

for each of the junction nodes

$$\sum_{n\in\mathbb{C}_{\text{IN}}}\left\{\frac{\partial}{\partial t}\left[\rho\left(p_{\text{Joint}}\ {}^{(n)}T\left\{{}^{(n)}Y_m\left|m=\overline{1N_s}\right.\right\}\right)\right]{}^{(n)}\Theta\right\}+$$

$$+\frac{\partial}{\partial t}\left[\rho\left(p_{\text{Joint}}\ T_{\text{Joint}}\left\{Y_{m\text{Joint}}\left|m=\overline{1N_s}\right.\right\}\right)\right]\sum_{n\in\mathbb{C}_{\text{OUT}}}{}^{(n)}\Theta+$$

$$+\sum_{n=1}^{N}\left\{\frac{\partial}{\partial x}\left[\rho\left(p_{\text{Joint}}\ {}^{(n)}T\left\{{}^{(n)}Y_m\left|m=\overline{1N_s}\right.\right\}\right){}^{(n)}w\right]{}^{(n)}\Theta\right\}=0$$

$${}^{(n)}\left(\frac{\partial}{\partial t}(\rho Y_m f)\right)+{}^{(n)}\left(\frac{\partial}{\partial x}(\rho Y_m wf)\right)=0m=\overline{1N_S-1}Y_{N_S}=1-\sum_{m=1}^{N_S-1}Y_m n\in\mathbb{C}_{\text{IN}}$$

$$Y_{m,\text{Joint}}=\sum_{n\in\mathbb{C}_{\text{IN}}}{}^{(n)}(\rho|w|fY_m)/\sum_{n\in\mathbb{C}_{\text{IN}}}{}^{(n)}(\rho|w|f)m=\overline{1N_S}$$

$${}^{(n)}\left(\frac{\partial}{\partial t}\left[\rho\left(\varepsilon+\frac{w^2}{2}\right)\right]\right)+{}^{(n)}\left(\frac{\partial}{\partial x}\left[\rho w\left(\varepsilon+\frac{w^2}{2}\right)\right]\right)=$$

$$=-{}^{(n)}\left(\frac{\partial(pw)}{\partial x}\right)-{}^{(n)}\left(\rho wg\frac{\partial z_1}{\partial x}\right)+{}^{(n)}Q-{}^{(n)}\left(\frac{\Phi\left({}^{(n)}T^{(n)}T_{\text{am}}\right)}{f}\right)n\in\mathbb{C}_{\text{IN}}$$

$$h_{\text{Joint}}=\sum_{n\in\mathbb{C}_{\text{IN}}}{}^{(n)}(\rho|w|hf)/\sum_{n\in\mathbb{C}_{\text{IN}}}{}^{(n)}(\rho|w|f)n\in\begin{cases}\mathbb{C}_{\text{IN}}\in\overline{1N}if^{(n)}(ws)\geq 0\\ \mathbb{C}_{\text{OUT}}\in\overline{1N}if^{(n)}(ws)0\end{cases}$$

$$T_{\text{Joint}}=T\left(p_{\text{Joint}}\ \varepsilon_{\text{Joint}}\left\{Y_{m\text{Joint}}\left|m=\overline{1N_s}\right.\right\}\right)\rho_{\text{Joint}}=\rho\left(p_{\text{Joint}}\ T_{\text{Joint}}\left\{Y_{m\text{Joint}}\left|m=\overline{1N_s}\right.\right\}\right)$$

$$\varepsilon_{\text{Joint}} = h_{\text{Joint}} - p_{\text{Joint}} / \rho_{\text{Joint}} \sum_{n \in C_{\text{IN}}} {}^{(n)}(\rho|w|f) = \sum_{n \in C_{\text{OUT}}} {}^{(n)}(\rho|w|f)$$

$${}^{(n)}p = p_{\text{Joint}}\; n = \overline{1N}\, {}^{(n)}\rho = \rho_{\text{Joint}}\; n \in \mathbb{C}_{\text{OUT}}\, {}^{(n)}T = T_{\text{Joint}}\; n \in \mathbb{C}_{\text{OUT}}\, {}^{(n)}\varepsilon = \varepsilon_{\text{Joint}}\; n \in \mathbb{C}_{\text{OUT}}$$

$${}^{(n)}Y_m = Y_{m,\text{Joint}}\; n \in \mathbb{C}_{\text{OUT}}\, m = \overline{1N_S}\, {}^{(n)}(z_1) = {}^{(\xi)}(z_1) \text{ for any} n\xi \in \overline{1N}$$

equation of state:

$$h = \varepsilon + p/\rho p = p\left(\rho T\left\{Y_m \middle| m = \overline{1N_s}\right\}\right)\rho = \rho\left(pT\left\{Y_m \middle| m = \overline{1N_s}\right\}\right)$$

$$\varepsilon = \varepsilon\left(pT\left\{Y_m \middle| m = \overline{1N_s}\right\}\right)T = T\left(p\varepsilon\left\{Y_m \middle| m = \overline{1N_s}\right\}\right)$$

where the subscript "Joint" means that the physical parameter of the gas mixture belongs to the pipeline junction i.e. to the volume ${}^{(0)}V$ h Joint is enthalpy of the gas mixture in the pipeline junction.

The numerical analysis of the mathematical models (1–12) and (1, 3, 7, 11, 13–24) under consideration will be carried out by hybrid algorithm. It comprises Integro-Interpolation Method by A. Tikhonov and A. Samarsky (IIM) (Russian analog of the Finite Volume Method) (Tikhonov & Samarsky, 1999) and Lagrangian Particle Method by Pryalov (LPM).

To illustrate the parametric classes used for the difference equations in IIM, it is possible to present the class of the difference equations for a mathematical model of the non-isothermal transient motion of a multi-component gas mixture through a GPS line (see (1–4, 12)) (Seleznev, 2007):

$$\left[(\rho F)^{(-S)}\right]_{\substack{t\\+}} + \left(\rho_{(-R)} w_{(0.5)} f_{(0.5)}\right)_{\substack{x\\+}}^{(\sigma\theta)} = 0$$

$$\left[(\rho F)^{(-S)} (Y_m)^{(-S)}\right]_{\substack{t\\+}} + \left(\rho_{(-R)} w_{(0.5)} f_{(0.5)} (Y_m)_{(-R)}\right)_{\substack{x\\+}}^{(\sigma\theta)} -$$
$$- \left[(\rho f D_m)_{(-R)} \delta(Y_m)^a\right]_{\substack{x\\+}}^{(\sigma\theta)} = 0m = \overline{1N_S - 1}Y_{N_S} = 1 - \sum_{m=1}^{N_S-1} Y_m$$

$$\left[(\rho F)^{(-S)} w^{(-S)}\right]_{\substack{t\\+}} + \left(\rho_{(-R)} w_{(0.5)} f_{(0.5)} w_{(-R)}\right)_{\substack{x\\+}}^{(\sigma\theta)} =$$
$$= -\left(B^-\gamma^- p_{\bar{x}} + B^+\gamma^+ p_x\right)^{(\sigma\theta)} - g\left[\rho\left(B^-\gamma^-(z_1)_{\bar{x}} + B^+\gamma^+(z_1)_x\right)\right]^{(\sigma\theta)} - \frac{\pi}{4}(\lambda\rho|w|r)^{(\sigma\theta)} w^{(\sigma\theta)}$$

$$\left[(\rho F)^{(-S)} \varepsilon^{(-S)}\right]_{\substack{t\\+}} + \left(\rho_{(-R)} w_{(0.5)} f_{(0.5)} \varepsilon_{(-R)}\right)_{\substack{x\\+}}^{(\sigma\theta)} + K_t\left(\rho F\frac{w^2}{2}\right) + K_x\left(\rho w f\frac{w^2}{2}\right)^{(\sigma\theta)} =$$
$$= -\left(p_{(-R)} w_{(0.5)} f_{(0.5)}\right)_{\substack{x\\+}}^{(\sigma\theta)} - g\left[\rho\left(B^-\gamma^-(z_1)_{\bar{x}} + B^+\gamma^+(z_1)_x\right)\right]^{(\sigma\theta)} w^{(\sigma\theta)} - p^{(\sigma\theta)}\left[F^{(-S)}\right]_{\substack{t\\+}} +$$
$$+ (QF)^{(\sigma\theta)} + \left[(kf)_{(-R)}\delta T^a\right]_{\substack{x\\+}}^{(\sigma\theta)} - \phi^{(\sigma\theta)} + \sum_{m=1}^{N_S}\left[(\rho D_m f)_{(-R)}(\varepsilon_m)_{(-R)}\delta(Y_m)^a\right]_{\substack{x\\+}}^{(\sigma\theta)}$$

$$\varepsilon_m = \varepsilon_m(\{S_{\text{mix}}\})m = \overline{1N_S}T_1 = T_2 = \ldots = T_{N_S} = Tp = p(\{S_{\text{mix}}\})\varepsilon = \varepsilon(\{S_{\text{mix}}\})$$

$$k = k(\{S_{\text{mix}}\})D_m = D_m(\{S_{\text{mix}}\})m = \overline{1N_S}$$

Where F B^+ B^- are the expressions approximating, fr is the expression approximating R (the ype of these expressions is defined upon selection of a particular scheme from the class of schemes); s^a

s^b r^a r^b are parameters of the class of schemes (e.g., by specifying s^a=s^b=1r^a=r^b=05, =1, =0, a two-layer scheme with central differences is selected from the class of scheme, and by specifying *sa*=*sb*=1*rarb* is according to the principles of "upwind" differencing, =1, =0 is a two-layer "upwind" scheme); K_t and K_x are the differential-difference operators of functions (0.5 Fw^2) and (0.5 wfw^2) over time and space, respectively (the type of these operators is defined upon selection of a particular scheme from the class of schemes); ϕ( ) is a difference expression approximating function (TT_{am}). The difference equations (25–30) are supplemented by difference expression of initial and boundary conditions, as well as of conjugation conditions.

To record the parametric class of the difference equations (25–30), we used notations of a non-uniform space-time mesh $\{x_i, t_j\}$, where x_i and t_j are coordinates of the mesh node numbered *i* over space, and *j*, over time, *i*, *j*∈*Z*, *Z* being a set of nonnegative integers.

To explain the notations, it is expedient to consider an individual computational cell containing the node $\{x_i, t_i\}$ (mesh $x = x_i^a$ $x = x_i^b$ $t = t_j^a$ and $t = t_j^b$ base node) and bounded by straight lines $x_{i-1} \le x_i^a \le x_i$ $x_i \le x_i^b \le x_{i+1}$ $t_{j-1} \le t_j^a \le t_j$ $t_j \le t_j^b \le t_{j+1}$ $x_i^a \ne x_i^b$ $t_j^a \ne t_j^b$ Let us introduce the so-called weighing parameters:

$$r_i^a = \frac{x_i^a - x_{i-1}}{x_i - x_{i-1}} = \frac{x_i^a - x_{i-1}}{h_i} r_i^b = \frac{x_i^b - x_i}{x_{i+1} - x_i} = \frac{x_i^b - x_i}{h_{i+1}} s_j^a = \frac{t_j^a - t_{j-1}}{t_j - t_{j-1}} = \frac{t_j^a - t_{j-1}}{\tau_j} s_j^b = \frac{t_j^b - t_j}{\tau_{j+1}}$$

where *hi*=*xi*−*xi*−1 and *hi*+1=*xi*+1−*xi* are steps "backward" and "forward" over the space coordinate for the *i* node; $_j$=t_j−t_j−1 and $_j$+1=t_j+1−t_j are steps "backward" and "forward" over the time coordinate on the *j* time layer; $_i$=h_i+1/*hi* and $_j$= $_j$+1/ $_j$ are mesh parameters characterizing non-uniformity of the space and time mesh. To describe mesh function *y*=*y*(*xt*) , the system (25–30) used the following notations:

$$y_i^j = y(x_i t_j) \, y_i = y_i(t) = y(x_i t) \, y^j = y^j(x) = y(x t_j)$$

Where there was applied the quadratic approximation

$$y(xt) = a_i^{y-}(t)x^2 + b_i^{y-}(t)x + c_i^{y-}(t)x \in [x_{i-1} x_i]$$
$$y(xt) = a_i^{y+}(t)x^2 + b_i^{y+}(t)x + c_i^{y+}(t)x \in [x_i x_{i+1}]$$

the system used the following notations:

$$\delta y(xt) = \frac{\partial y}{\partial x} = 2a_i^{y-}(t)x + b_i^{y-}(t)x \in [x_{i-1} x_i]\delta y(xt) = \frac{\partial y}{\partial x} = 2a_i^{y+}(t)x + b_i^{y+}(t)x \in [x_i x_{i+1}]$$
$$(\delta y^a)_i^j = 2a_i^{y-}(t_j)x^a + b_i^{y-}(t_j)\left(\delta y^b\right)_i^j = 2a_i^{y+}(t_j)x^b + b_i^{y+}(t_j)$$

Also, the system (25–30) used the following index-free notations:

$$y = y_i^j h = h_i \alpha = \alpha_i \tau = \tau_j \beta = \beta_j x = x_i t = t_j \widehat{y} = y_i^{j+1} \breve{y} = y_i^{j-1} y(+1) = y_{i+1}^j$$
$$y(-1) = y_{i-1}^j r^a = r_i^a r^b = r_i^b s^a = s_j^a s^b = s_j^b ij \in Z y^{(a)} = ay + (1-a)\breve{y}$$
$$y_{(b)} = by + (1-b)y(-1) y^{(ab)} = a\widehat{y} + (1-a-b)y + b\breve{y}\, y^{(-S)} = s^a y + (1-s^a)\breve{y}$$
$$y^{(+S)} = s^b \widehat{y} + (1-s^b)y y_{(-R)} = r^a y + (1-r^a)y(-1) y_{(+R)} = r^b \cdot y(+1) + \left(1-r^b\right)y$$
$$y_{(-R)}(+1) = y_{(+R)} y_{(+R)}(-1) = y_{(-R)} \widehat{y}^{(-S)} = y^{(+S)} \breve{y}^{(+S)} = y^{(-S)} \Delta t^- = t - t^a = (1-s^a)\tau$$
$$\Delta t^+ = t^b - t = s^b \widehat{\tau} = \beta s^b \tau \Delta x^- = x - x^a = (1-r^a)h \Delta x^+ = x^b - x = r^b h(+1) = \alpha r^b h$$
$$\Delta t = t^b - t^a = \Delta t^+ + \Delta t^- = (1-s^a)\tau + s^b\widehat{\tau} = (1-s^a)\tau + \beta s^b \tau$$
$$\Delta x = x^b - x^a = \Delta x^+ + \Delta x^- = (1-r^a)h + r^b h(+1) = (1-r^a)h + \alpha r^b h$$
$$\gamma^- = \Delta x^- / \Delta x = (1-r^a)h / \left[(1-r^a)h + r^b h(+1)\right] = (1-r^a) / \left[(1-r^a) + \alpha \cdot r^b\right]$$
$$\gamma^+ = \Delta x^+ / \Delta x = r^b h(+1) / \left[(1-r^a)h + r^b h(+1)\right] = \alpha \cdot r^b / \left[(1-r^a) + \alpha \cdot r^b\right]$$
$$\delta y^a(+1) = \delta y^b \delta y^b(-1) = \delta y^a y_x = [y(+1) - y]/h(+1) y_{\bar{x}} = [y - y(-1)]/h$$
$$y_{\bar{x}}(+1) = y_x y_x(-1) = y_{\bar{x}} y_{\underset{+}{t}} = \left[\widehat{y} - y\right]/\Delta t y_{\underset{+}{x}} = [y(+1) - y]/\Delta x$$

LPM is illustrated by the example of solution of energy equation from gas dynamics equations set for a single-component gas transmitted through an unbranched pipeline. Imaginary Lagrange particles are distributed along the pipeline. They are considered weightless. This allows them to move together with the fluid. Due to the small size, each particle can instantaneously acquire the temperature of the ambient fluid. Thus, by tracking the motion of such Lagrange particles along with the fluid and their temperature, one can analyze the process of heat transfer through multi-line GPSs. Energy equation is easy to derive by simplifying and transforming equation (4) accounting for (1-3), (12) and (23). The simplified equation will have the following form:

$$\frac{\partial h}{\partial t}+w\frac{\partial h}{\partial x}=\frac{w}{\rho}\frac{\partial p}{\partial x}+\frac{1}{\rho}\frac{\partial p}{\partial t}+\frac{\lambda|w|^3}{4R}-\frac{\Phi(TT_{am})}{f\rho}\text{ or }\frac{Dh}{Dt}=G(xtT(h))$$

Where

$$G(xtT)=\frac{w}{\rho}\frac{\partial p}{\partial x}+\frac{1}{\rho}\frac{\partial p}{\partial t}+\frac{\lambda|w|^3}{4R}-\frac{\Phi(TT_{am})}{f\rho}$$

$D\xi/Dt=\partial/\partial t+w\partial/\partial x$ is a derivative of an arbitrary function over t in the direction

$$dx/dt=w(xt)$$

This direction is called characteristic, and the equation is called the equation of characteristic direction. The second equation in (31) is called the characteristic form of the first equation in (31) or the differential characteristic relation. From the physical standpoint, the derivative Dh/Dt corresponds to the substantial derivative, and the solution of equation (33) defines the coordinate of the continuum particle (in our case, the spatial coordinate of the fluid flow cross section) at each time.

Considering the known thermodynamic relationship

$$dh=c_p dT-\mu c_p dp$$

where cp is specific heat capacity at constant pressure; $\mu$is the Joule-Thomson factor, equation (31) can be transformed as follows:

$$\frac{DT}{Dt} = \left(\mu + \frac{1}{\rho c_p}\right)\frac{Dp}{Dt} + \frac{\lambda|w|^3}{4Rc_p} - \frac{\Phi(TT_{\mathrm{am}})}{f\rho c_p}$$

Equation (35) is satisfied along each characteristic curve (33). These curves per se describe the trajectory of fluid particles. In other words, these equations describe the change in the fluid temperature for each cross section of the transported product flow.

When implementing the LPM, fluid flow parameters (such as pressure and velocity) are obtained using a difference scheme, while the gas temperature distribution is obtained based on the analysis of the Lagrange particle motion. For each particle, equation (35) is solved. The form of this equation enables such simulations, because it corresponds to the change in time of the temperature at each cross section of the fluid flow. Within this problem statement (which is Lagrangian with respect to each particle), equation (35) transforms into an ordinary differential equation (ODE) relative to the marching variable:

$$\frac{dT}{dt} - \left(\mu + \frac{1}{\rho c_p}\right)\frac{dp}{dt} = \frac{\lambda|w|^3}{4Rc_p} - \frac{\Phi(TT_{\mathrm{am}})}{f\rho c_p}$$

Numerical analysis of ODE (36) can be carried out using different ODE solution procedures, for example, the known Runge - Kutta method with an adjustable accuracy of solution. As initial temperature of Lagrange particles one uses respective values from the defined initial conditions (i.e., for each Lagrange particle, its temperature is assumed equal to the fluid temperature at the location of the given particle).

As particles move together with the fluid flow towards the outlet pipe boundary, one needs to introduce new Lagrange particles at the inlet boundary at some regular intervals. The initial temperature of each introduced particle should be defined based on the boundary conditions related to the inlet boundary of the given pipe. The Lagrange particles that leave the pipe are deleted. As applied to the inlet boundaries of the outlet pipelines of each junction node, the temperature of the introduced particles should be defined in accordance with equations (10) and (12).

Since LPM for solving the energy equation is not related immediately to the finite difference mesh used to solve the continuity and momentum equations, this mesh has almost no effect on the accuracy of the method proposed. Thus, high-accuracy computed values of gas temperature are obtained without mesh refinement, and this leads to a considerable speedup of computations.

Also, as there is no direct relationship between LMP and the finite difference mesh, the method is free of artificial viscosity and computational dispersion (Fletcher, 1988). As a result, LPM yields solutions without "scheme smoothing" of temperature fronts, which is consistent with actual physical processes. This makes such simulations more credible than the simulations, in which the energy equation is solved using difference schemes.

SIMULATION OF A CS BY CFD-SIMULATOR

The principal task of mathematical simulation of stable and safe operation of a CFS is to determine physical parameters of gas at the CFS outlet on the basis of the known values of gas flow parameters at the CFS inlet. To construct a 1D mathematical model of a CFS in a CFD-simulator, we used a well-known polytropic model of a CFS developed by A. Stepanov. The model is based on the combination of analytical dependencies for polytropic fluid dynamics processes and empirical characteristics obtained for each CFS during its full-scale testing.

When simulating steady modes of CS operation, an isothermal model is used for description of the gas flow in a CGP and DC, and an isobaric model - for description of the gas flow in an ACD. The power drive is simulated by specifying an analytical dependency of the capacity at the CFS shaft on energy expenditures. Such approach provides for the simplicity of the conjugation of models and a high, from the practical standpoint, veracity of simulation.

As was noted above, a CFD-simulator of a particular CS is a result of combining GCU, ACD and DC models, by application of CGP models, into a single integral network model of the CS in accordance with the process flow charts of the actual CS (fig. 2).

As proposed by V. Seleznev (Kiselev et al., 2005), in order to determine parameters of steady modes of natural gas transmission through a CS, generally, it is necessary to solve a system of nonlinear algebraic equalities under simple constraints on unknown variables. The system includes the law of conservation of gas masses at CGP branching points and one of the group of equations representing either conditions for conservation of mass flow rate at inlet and outlet CGPs in one branch, or conditions for equality of natural gas heads in parallel branches, where a branch is a segment of a pipeline system, which comprises an inflow (inlet) CGP, CFS and an outflow (outlet) CGP (see fig. 2).

As independent decision variables we used fractions of a mass flow rate of natural gas transmitted through separate branches of a CS, ratios of compression by compressor shops and ratios of compression by GCUs working as the first stage of transmitted gas compression at compressor shops. Such a set of variables allowed to reduce the problem dimension and narrow the range of search for problem solution, by more accurate specification of constraints on variables. This allowed to considerably save running time.

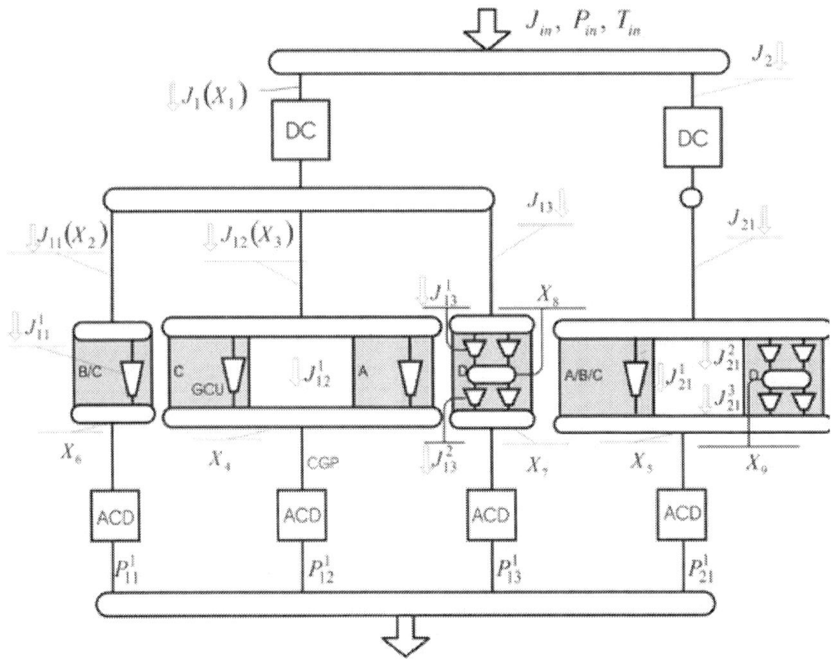

Figure 2: The scheme of decision independent variables assignment for on-line technological analysis of gas transmission through CS

The mathematical model for the CS scheme presented in fig. 2 can be written as follows:

$$\begin{cases} P_{11}^1 (J_{in} P_{in} T_{in} J_1(X_1) J_{11}(X_2) X_6) - P_{12}^1 (J_{in} P_{in} T_{in} J_1(X_1) J_{12}(X_3) X_4) = 0 \\ P_{12}^1 (J_{in} P_{in} T_{in} J_1(X_1) J_{12}(X_3) X_4) - P_{13}^1 (J_{in} P_{in} T_{in} J_1(X_1) J_{13} X_8 X_7) = 0 \\ P_{13}^1 (J_{in} P_{in} T_{in} J_1(X_1) J_{13} X_8 X_7) - P_{21}^1 (J_{in} P_{in} T_{in} J_2 J_{21} X_9 X_5) = 0 \\ J_{11}(X_2) - J_{11}^1(X_6) = 0 J_{12}(X_3) - J_{12}^1(X_4) = 0 J_{13} - J_{13}^1(X_8) = 0 \\ J_{13}^1(X_8) - J_{13}^2(X_8 X_7) = 0 J_{21} - J_{21}^1(X_5) - J_{21}^2(X_9) = 0 J_{21}^2(X_9) - J_{21}^3(X_9 X_5) = 0 \\ 0 X_i 1 i = \overline{13} \varepsilon_{\max}^1 X_j \varepsilon_{\max}^{shop} j = \overline{47} \varepsilon_{\min}^1 X_k \varepsilon_{\max}^1 k = \overline{89} \end{cases}$$

where$P111$, $P112$, $P113$, $P121$ are the natural gas pressure at the outlet of each of the CS branches, $JinPinTin$ are the natural gas flow rate, pressure and temperature at the CS inlet, $J1$, $J2$ are the natural gas mass flow rates through "branches I", $J11$, $J12$, $J13$, $J21$,are the natural gas mass flow rates through "branches II", $X6$, $X4$, $X7$, $X5$ are the ratios of compression for compressor shops, *X*8, *X*9 are the ratios of compression for GCU groups of the first stages of compressor shops, 1minmax is a minimal/maximal ratio of compression for GCU groups of the first stage, *shop*max is the maximal ratio of compression for compressor shops.

To assure a safe mode of CS operation, it is required to observe the following restrictions on: maximal volumetric capacity *Qj* of each operating CFS; frequency *uj* of the CFS shaft rotation; maximal capacity *Nj* of the CFS drive; maximal outlet pressure *Pj* of the CFS, which is determined by the pipe›s strength; maximal temperature *Tj* at the CFS outlet, which is determined by the insulating coating; the lower value of pressure at the outlet of each CFS, related to the requirements to maintain pressure at major gas tapping and boundary gas pipeline points. These restrictions can be formulated as one-sided and two-sided weak inequality:

$$\left\{Q_j^{\min} \le Q_j(\vec{X}) \le Q_j^{\max}\ u_j^{\min} \le u_j \le u_j^{\max}\ N_j(\vec{X}) \le N_j^{\max}\ P_j^{\min} \le P_j(\vec{X}) \le P_j^{\max}\ T_j(\vec{X}) \le T_j^{\max}\right\}$$

where index *j*=1...*M* means the number of CFS. Also, it is necessary to comply with restrictions on: positions of operational points at the CFS performance curves, related to surge-free operation requirements; conditions related to CFS drive's stable operation; etc. Compliance with the latest restrictions can only be analyzed in the process of a detailed simulation of the mode of gas transmission through the CS.

The system of nonlinear algebraic inequalities (38) describes technological, operating and design constraints on CS equipment operation. When formulating the constraints in a CFD-simulator, a maximum consideration is given to the specifics of the CS process flow chart and the modes of its operation, including a possibility of surging in "CFS - Adjacent CGP" (Seleznev & Pryalov, 2007).

The resultant system (37, 38) represents a generalized system of nonlinear algebraic equalities and inequalities. One of the methods to solve the system of nonlinear algebraic equalities and inequalities is the well-known Interior Point Method, which, in a CFD-simulator, is implemented by statement and solution of an equivalent problem of mathematical optimization (Dennis & Schnabel, 1988). The equivalent optimization problem is solved by the method of modified Lagrange functions. If no solution can be found, the possibility of failure occurrence is admitted. The method we use makes it possible to identify constraints which are not complied with and to evaluate the extent of such non-compliance.

SIMULATION OF A GTN BY CFD-SIMULATOR

Industrial application of CFD-simulators in the gas industry is shown by the example of two acute problems: cutting down the power costs for natural gas transportation through pipeline networks; finding out the reasons for discrepancy between estimates of supplied gas volume by seller and consumer. To resolve these problems, one should be able to simulate the processes of GTN operation in a credible way. Let us consider the GTN simulation algorithm of S. Pryalov as applied to a fragment of a hypothetical CS (fig. 3).

In this CS fragment, three GCUs are combined in parallel into a group by means of a CGP. Fig. 3a uses the following notations:*PT*– pressure and temperature of the gas (the double number subscripts indicate that the physical parameters of the natural gas flow are attributed to the CFS inlet or outlet with a respective number; parameters at the CS inlet are denoted with the lower index *inJii*=123– gas mass flow in the CS pipeline branch with its own number;*nii*=123– speed of the *i*-th CFS shaft;ξ– friction factor for CGP (the double number subscripts indicate that the quantity is attributed to a certain CGP).

Let us conventionally isolate one (the second from the top in fig. 3a) CS branch (fig. 3b): between points **A** and**B**. In this case, it is assumed that respective boundary conditions are defined at the branch inlet and outlet points. Such boundary conditions can include: pressure *P*A and temperature *T*A at the branch inlet, and pressure *P*B and heat flux *q*B (or temperature*TB*) at the branch outlet (note that generally speaking these quantities are functions of time).

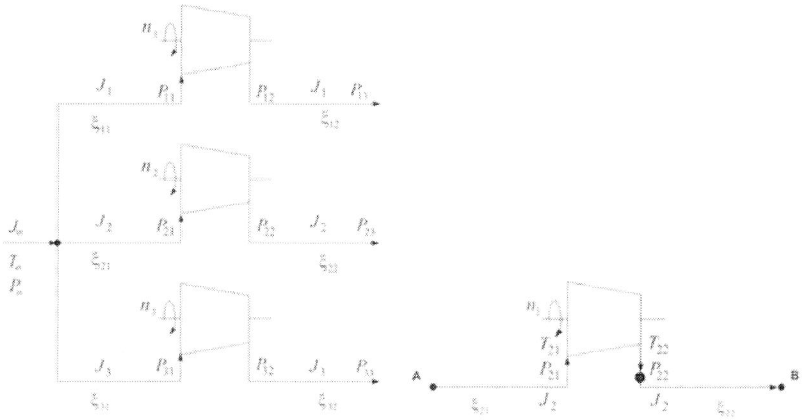

a) b)

Figure 3 A diagram of a hypothetical CS (a) and a diagram for the branch point A to point B (b)

The value of the mass flow rate through the CFS *J*=*J*CFS2=*J*CFS2(*t*) in this case is a solution of a non-linear algebraic equation (lower indices of pressures are the same as in fig. 3):

$$\Psi_2(J) = P_{22}^{CFS}(J) - P_{22}^{CGP}(J) = 0$$

where *P*CFS22(*J*)= 2(*P*CGS21(*J*)*T*CGS21(*J*)*J*)·*P*CGS21(*J*) is pressure at the CFS outlet; 2(*P*CGS21(*J*)*T*CGS21(*J*)*J*)is gas compression rate produced by the CFS and computed in the CFD-simulator using a semi-empirical CFS model (Seleznev & Pryalov, 2007); *P*CGS21(*J*) and *T*CGS21(*J*) are the outlet pressure and temperature of the inlet CGP from the simulations of the gas flow parameters in the given pipeline, in which boundary conditions at its outlet include the target mass flow *J* and heat flux conservation at the outlet of the inlet CGP; *P*CGS22(*J*)=*P*CGS22(*J**T*CFS22(*J*))is inlet pressure of the inlet CGP from the simulations of the gas flow parameters in the given pipeline, in which boundary conditions at its outlet include the target mass flow *J* and gas temperature equal to the gas temperature *T*CFS22(*J*) at the CFS outlet.

The functions in this equation are determined numerically by using the CFD-simulator at each step of the iterative solution of equation (39). Therefore, the target distributions of the fluid dynamics parameters of both CGPs and operating parameters of the GCU are obtained automatically in the course of solving equation (39).

Equation (39) in the CFD-simulator is solved numerically using the well-known modified Newton procedure, in which equation (39) is solved by iterations using a finite-difference Jacobian approximation. In the paper (Dennis & Schnabel, 1988) the authors show that this method preserves its q-quadratic convergence of the classical Newton procedure for solving non-linear algebraic equations provided that the finite-difference step length is chosen properly.

For simulations of transient CS operation conditions, the computational model of a hypothetical CS fragment (see fig.3a) will comprise sub-models of an inlet CGP, three CFSs and an outlet CGP. The inlet CGP includes pipelines from the point of transported gas flow entry into the CS to the points of the gas flow entry into each CFS. The outlet CGP includes pipelines from the points of gas flow exit from each CFS to the point of gas flow exit from the CS. These inlet and outlet CGPs in this case are branched, multi-line pipelines, which are simulated by models (1–12) or (1, 3, 7, 11, 13–24). It is assumed that boundary conditions at the transported gas flow inlet/outlet points of the CS are defined in accordance with the principles described above. Here, the values of the mass flow rate through each CFS, $JCFSi=JCFSi(t)$ where$i=\overline{1N}$ N is the number of CFSs in the group), constitute a solution to the set of non-linear algebraic equations:

$$\Psi_i(J_i) = P_{i2}^{CGP}(J_i) - P_{i2}^{CFS}(J_i) = 0 i = \overline{1N}$$

The notations used in (40) are similar to those used in equation (39). The subscripts of pressures correspond to the indices shown in fig. 3.

The target distributions of the fluid dynamics parameters of the CGPs and operating parameters of the GCU equipment are obtained automatically by numerical simulation of equations (40) (as a result of using the CFD-simulator for solving this set of equations to determine the values of its constituent functions).

Numerical solution of equations (40) in industrial application of the CFD-simulator is performed by known quasi-Newton methods. In the first line, one can recommend using the modified Broyden method (Dennis & Schnabel, 1988) as one of the best performing extensions of the classical secant method to the case of numerical solution of non-linear algebraic equations. This method can have q-superlinear local convergence with an r-order of212N/

It is not difficult to extend the above method to the whole GTN. In this case, equations (40) instead of the mass flow rates through the CFSs will contain the mass flow rates through the CS as a whole. By analogy with simulations of gas transportation through an individual CS, due to the use of the CFD-simulator, target distributions of parameters for the whole multi-line GPS and operating parameters of GCU, CFS, DC and ACD are obtained automatically by numerical simulation of modified equations (40) by the quasi-Newton methods.

OPTIMIZATION OF GAS TRANSMISSION EXPENDITURES USING CFD-SIMULATOR

To avoid too many technical details without loss of generality, here we consider GTNs with serially connected CSs (fig. 4).

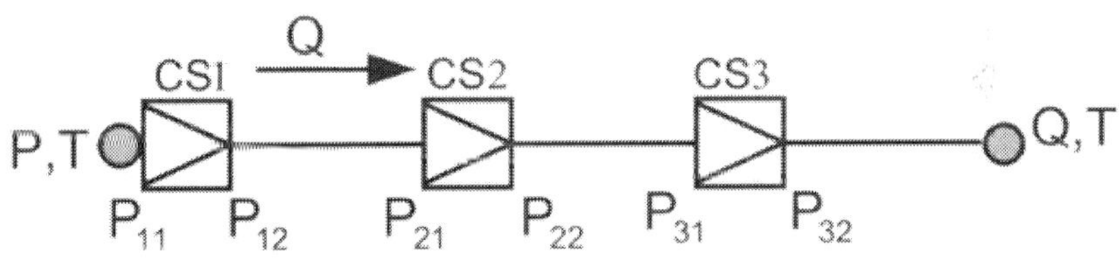

Figure 4: A diagram of a hypothetical GTN of serially connected compressor stations

Fig. 4 uses the following notations:PTQ– pressure, temperature and volumetric flow rate. The symbols CS1, CS2 and CS3 in fig. 4 are used in the diagram to designate CSs with their serial number. The double number subscripts indicate that the physical parameters of the natural gas flow are associated with the inlet or outlet of the CS with a respective serial number. The horizontal lines between the CSs in the diagram show multi-line GPSs. The flow direction in fig. 4 is from left to right. The GTN diagram is typical of many gas transportation companies. As an example one can mention GTNs of Tomsktransgas, Russia, and SPP, a.s., Slovakia.

The energy criterion of reduced expenses for failure-free natural gas transportation through GTNs has the following form: One should obtain calculated estimates of GTN equipment controls corresponding to minimum costs of integral energy costs related to transient conditions of gas transportation through the GTN in a given time interval subject to compliance with gas specifications at GTN control points and, at the same time, observance of applicable restrictions to ensure safe and environmentally sound operation of the GTN.

In such a criterion, estimates of the cost function for integral energy expenditures are made using the CFD-simulator. For this purpose, the CFD-simulator features functional dependencies of energy carrier consumption in GCUs on the power consumption at CFS shafts for all GCUs of the GTN. Note that plotting distributions of energy carrier consumption in GCUs over time in the CFD-simulator belongs to GTN performance monitoring procedures and is one of its standard functions (Seleznev & Pryalov, 2007). If the energy carrier consumption and the price of these energy carriers are known, it will be easy for a gas transportation company to calculate their total cost.

The requirement of compliance with gas specifications at GTN control points within the criterion is consistent with the requirement for gas transportation companies to fulfill their contractual obligations. The requirement of compliance with restrictions to ensure GTN safety and environmental soundness comes to the observance of process, performance, design and other restrictions as applied to the operation of the GTN equipment. These restrictions can be formalized mathematically as a set of simple constraints on the target values of the GTN equipment controls and a set of non-linear weak two-sided (and/or one-sided) algebraic inequalities. Function values in these inequalities (GTN equipment operating data and fluid dynamics parameters of the gas flow in the network) are calculated using the CFD-simulator.

In practice, in implementing the criterion mathematically, one usually seeks a local minimum of the function of integral expenses subject to given constraints. This is primarily related to the complexity (and, as a rule, polymodality) of the target function, complexity of the constraint functions and uncertainties in input data coming to the CFD-simulator from SCADA systems.

An optimization algorithm for transient GTN operation conditions was developed by V. Seleznev in 2003. It provides for preliminary predictive simulations, using the CFD-simulator, of physical parameters of the gas flow through the GTN for a given time interval based on the predefined history of GTN operation and gas flow parameters at the GTN boundaries. The value of the function of energy (or financial) expenses for gas transportation through the GTN, which is related immediately to the operating parameters of its equipment and gas flow parameters, is calculated at each time step of this simulation. Thus, during the predictive simulations, the time dependence of expenses is automatically generated for a so-called non-optimized prediction.

Setting up an optimized prediction for the given time interval in a general case consists in the definition of time laws for the GTN equipment controls to reduce the costs of natural gas transportation. Target controls should first of all include regulation of the CFS shaft speed in the whole GTN. The speed regulation for all CFSs can result in shutdowns or starts of individual GCUs. The list of involved GCUs, including valve status data, are called the CS configuration (or configuration of the GTN as a whole).

As has been already mentioned above, energy (or financial) expenses at a given time point t depend on the status $\vec{\mathbf{u}}(t)$ of the whole package of GTN gas pumping equipment. These expenses will be represented in the form of a non-linear algebraic cost function,$Z(t\vec{\mathbf{u}}(t))$. It is evident that total expenses for the prediction period $[0T]$ will be an integral, $\int 0TZ(t\vec{\mathbf{u}}(t))dt$. From the mathematical viewpoint, minimizing the costs of natural gas transportation through the GTN over the predicted period comes to the minimization of the area of the 2D figure $\mathbf{\Phi}$ shown in fig. 5.

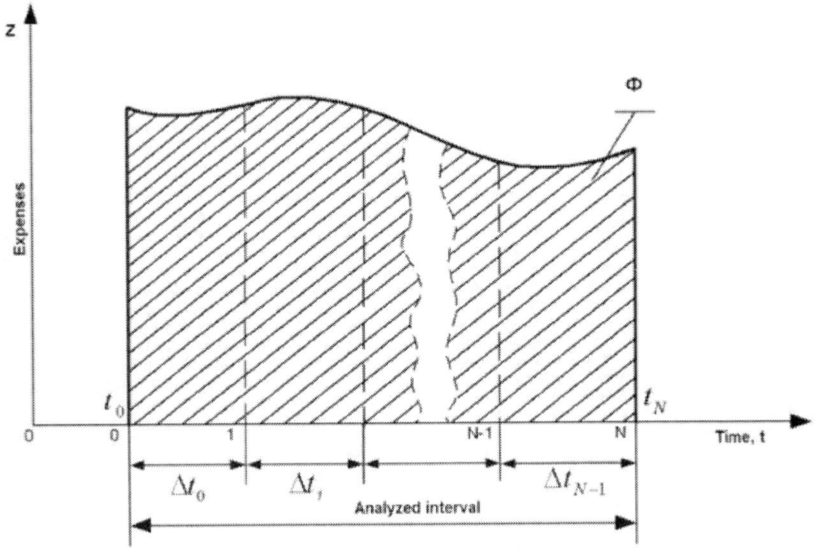

Figure 5: Initial function of gas transportation expenses in the analyzed interval (prediction period)

Mathematically, this can be represented in the following way:

$$J\left(\vec{\mathbf{u}}\right) = \int_0^T Z\left(t\vec{\mathbf{u}}(t)\right)dt \to \min t \in [0T]\vec{\mathbf{u}}(t) \in U(t) \subset R^m$$

Given that

$$\vec{\mathbf{W}}\left(t\vec{\mathbf{u}}(t)\right) \in \Omega = \left\{\vec{\mathbf{W}} \in R^k : \left(w_{\min}^{specified}\right)_j \leq w_j\left(t\vec{\mathbf{u}}(t)\right) \leq \left(w_{\max}^{specified}\right)_j j = \overline{1l}\right.$$
$$\left. w_j\left(t\vec{\mathbf{u}}(t)\right) \leq \left(w_{\max}^{specified}\right)_j j = \overline{l+1k}\right\}$$

where $U(t)$ is a given set to define the type of constraints related to the limited availability of control resources (simple constraints on the control quantities); $m$is the number of controls used for optimization; $\mathbf{W}^{\rightarrow}$ $(t\mathbf{u}^{\rightarrow}$ $(t))=(w1(t\mathbf{u}^{\rightarrow}$ $(t))...wk(t\mathbf{u}^{\rightarrow}$ $(t)))T$ is a given algebraic constraint vector-function; $k$is the total number of constraints in the form of two-sided and one-sided weak algebraic inequalities used for optimization; $\vec{\mathbf{w}}\ specified\mathrm{min}\in Rl$ and $\vec{\mathbf{w}}\ specified\mathrm{max}\in Rk$ are given vectors of the minimum and maximum permissible values to implement the constraints; $l$is the number of constraints in the form of two-sided weak algebraic inequalities used for optimization; $Rm$is the Euclidean m-dimensional vector space; $Rk$is the Euclidean k-dimensional vector space; $Rl$is the Euclidean l-dimensional vector space.

The first l rows of the vector-function $\vec{\mathbf{W}}\ (t\vec{\mathbf{u}}\ (t))$ describe process, design, environmental and other constraints on the GTN operating conditions. The rest of the constraint vector-function rows provide stabilization of the programmed status and GTN operating conditions. Stabilization of a given programmed status and GTN operating conditions here means maintaining true paths of GTN operating and gas flow parameters in some neighborhood of expected paths. Examples of such GTN parameters are gas pressures at the GTN boundaries, which should be strictly controlled in accordance with contractual obligations of the gas company. Thus, problem (41, 42) includes additional conditions, $wj(t\vec{\mathbf{u}}\ (t))\leq(wspecified\mathrm{max})\ jj=l+1k$, which will constrain the deviation of computed paths of specified parameters from their programmed paths. Algebraic functions in these constraints are generally modules of discrepancies between computed and programmed paths.

The technology of natural gas transportation through GTNs provides for stepwise variation of the GTN equipment controls (for example, individual GCUs or their groups) in time. Therefore, in setting up an optimal prediction of expenses, one should map a grid $t0t1\ldots tN$ onto the analyzed interval $[0T]$ (see fig. 5) in such a way as to provide constant control of the GTN equipment within the time interval $(titi+1)i=0N-1$, (fig. 6a), i.e:$\vec{\mathbf{u}}\ (t)=\vec{\mathbf{u}}\ i=const t\in(titi+1)\vec{\mathbf{u}}\ i\in U(t)$. To simplify the problem definition (without loss of generality), we assume conventionally that the step $t=ti+1-ti$, $i=0\ldots N-1$, is constant all over the prediction period.

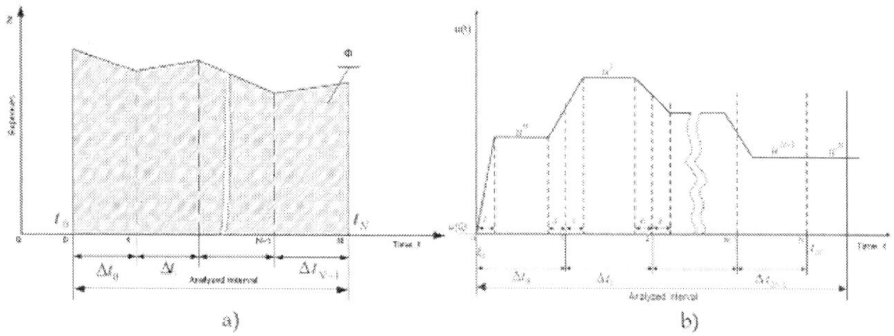

a) b)

Figure 6: Approximation sequence for the cost function

Given this, the function of the GTN equipment status will be represented as follows (fig.6a):

$$u_j(t) = \begin{cases} \frac{t - t_0}{\delta}\left(u_j^0 - u_j(0)\right) + u_j(0) t \in [0\delta) \\ u_j^i t \in [t_i + \delta t_{i+1} - \delta) \\ \frac{t - (t_{i+1} - \delta)}{2\delta}\left(u_j^{i+1} - u_j^i\right) + u_j^i t \in [t_{i+1} - \delta t_{i+1} + \delta) \end{cases}$$

Where i=0...N−1, j=1...m, uj(0) is a known value of the initial control at the initial time, uNj=uN−1j; the spacing is chosen empirically. In industrial applications the function of the GTN equipment status is often a function of shaft speed variation for all CFSs.

It is difficult to calculate the integral of expenses (41) in the constructed optimization model (41–43) analytically. Therefore, this integral is calculated numerically. The widely known method of trapezoids is considered quite acceptable as applied to this case, i.e. (fig.6b):

$$J(\vec{u}) = \int_0^T Z\left(t\vec{u}(t)\right)dt \approx 0.5\Delta t Z_0\left[\vec{u}(t_0)\right] + \Delta t \sum_{i=1}^{N-1} Z_i\left[\vec{u}(t_0 + i\cdot\Delta t)\right] + 0.5\Delta t Z_N\left[\vec{u}(t_N)\right]$$

Thus, for some set of discrete controls (**u**→ 0**u**→ 1...**u**→ N−1)T, formula (44) allows calculating the uality functional ($\vec{\mathbf{u}}$ 0$\vec{\mathbf{u}}$ 1... $\vec{\mathbf{u}}$ $N-1$). Hence, the problem of finding an optimal control (41–43) can be replaced with an equivalent problem of nonlinear programming with respect to independent vectors ($\vec{\mathbf{u}}$ 0$\vec{\mathbf{u}}$ 1...$\vec{\mathbf{u}}$ $N-1)T$

$$\tilde{J}\left(\vec{\mathbf{u}}^0\vec{\mathbf{u}}^1\ldots\vec{\mathbf{u}}^{N-1}\right) = 0.5Z_0\left[\vec{\mathbf{u}}^0\right] + \sum_{i=1}^{N-1} Z_i\left[\vec{\mathbf{u}}^i\right] + 0.5Z_N\left[\vec{\mathbf{u}}^{N-1}\right] \to \min$$

subject to the conditions (considering (43)):

$$\vec{\mathbf{u}}^i \in U = \left\{\vec{\mathbf{u}} \in R^m : \max\left[u^{i-1} - \eta_{constr}u_{\min}^{specified}\right]_j \le u_j^i \le \min\left[u^{i-1} + \eta_{constr}u_{\max}^{specified}\right]_j, j = \overline{1m}\right\}$$
$$i = \overline{1N-1}$$

$$\vec{\mathbf{W}}_i\left(\vec{\mathbf{u}}^0\vec{\mathbf{u}}^1\ldots\vec{\mathbf{u}}^{N-1}\right) \in \Omega = \left\{\vec{\mathbf{W}}_i \in R^k : \left(w_{\min}^{specified}\right)_j \le w_j^i\left(\vec{\mathbf{u}}^0\vec{\mathbf{u}}^1\ldots\vec{\mathbf{u}}^{N-1}\right) \le \left(w_{\max}^{specified}\right)_j\right.$$
$$\left. j = \overline{1l} w_j^i\left(\vec{\mathbf{u}}^0\vec{\mathbf{u}}^1\ldots\vec{\mathbf{u}}^{N-1}\right) \le \left(w_{\max}^{specified}\right)_j, j = \overline{l+1k}\right\} i = \overline{1N}$$

where $\vec{\mathbf{u}}$ *specified*min∈Rm and $\vec{\mathbf{u}}$ *specified*max∈Rm are given vectors of the minimum and maximum values of control parameters (as a rule, these vectors are time-independent); $\vec{\eta}$ *constr*∈Rm is a given constraint on the maximum step of a single j-th control change in time. The numerous constrains in (47) are attributed to the requirement that variation of GTN operating parameters on transition from one time layer to the next one should be smooth. One should be reminded here that calculations of expenses Zi($\vec{\mathbf{u}}$ i) for gas transportation through the GTN are enabled by the CFD-simulator.

In industrial applications for optimal predictions, knowing the optimal control path is more critical than knowing the minimum value of the quality functional. Therefore, it is reasonable to transform the problem statement (45) in the following way:

$$\widehat{J}\left(\vec{\mathrm{u}}^{1}\ldots\vec{\mathrm{u}}^{N-1}\right)=\sum_{i=1}^{N}Z_i\left(\vec{\mathrm{u}}^{i}\right)\to \mathrm{min where} \vec{\mathrm{u}}^{N}=\vec{\mathrm{u}}^{N-1}$$

To solve the problem of general non-linear programming (46–48), we use the method of modified Lagrange functions (Vasilyev, 2002). In accordance with this method, the modified target function of the optimized problem being solved is a sum of initial target function (48) and formalized constrains (47) weighted in a special way (by means of Lagrange multipliers and penalties). An equivalent task of searching for the minimum value of the modified target function subject to simple constraints on variables (46) is fulfilled using modified varied metrics algorithms (Vasilyev, 2002) or modified conjugate gradient algorithms (Vasilyev, 2002), which are resistant to the error accumulation in the course of arithmetic operations.

If the choice of the GTN configuration is considered as an additional control, one will have to introduce additional constraints into the optimization problem to represent process bans on frequent changes of GTN configurations during gas transportation. The choice of the GTN configuration does not require fundamental changes in the optimization approach or algorithm for transient conditions of gas transportation through the GTN. The above approach to the construction of optimal prediction for gas transportation through the GTN can be easily extended to the CS or group of GCUs.

NUMERICAL MONITORING OF GAS DISTRIBUTION DISCREPANCY USING CFD-SIMULATOR

Numerical monitoring of the discrepancy is based on a statement (for a specified time gap) and numerical solution of identification problem of a physically proved quasi-steady gas dynamics mode of natural gas transmission through specified gas distribution networks.

In large communities, natural gas is supplied to the consumers using medium or low pressure ring mains, being several dozen kilometers long. Gas from the supplier is transmitted to such mains through a GTN after its pressure is reduced by means of a system of gas reducers installed at inlet gas distribution stations (GDSs) (fig. 7). Major parameters of gas supplied by the gas transportation company to the seller are also measured at the GDS outlets. Here, major parameters of natural gas include its flow rate, pressure and temperature.

Gas from inlet GDSs is delivered to the ring main via the CGP network of the gas seller. Consumers receive gas from the ring mains through outlet CGPs leading from the ring main to the consumer. The length of the CGPs can range from several hundred meters to several kilometers. In the first approximation, each consumer is considered independent and provided with gas through one CGP, which is completely associated with the consumer (called "associated CGP" as the text goes). Consumer independence means that the consumer's gas cannot be delivered to other consumers.

Thus, the gas distribution network (GDN) under consideration comprises inlet CGPs from inlet GDSs, a ring main and associated CGPs. In fig. 7, the GDN under consideration is shown with gray color.

If the GDN operates properly, the seller seeks to sell the whole amount of gas received from the supplier. An exception in this case is natural gas forcedly accumulated in the GDN.

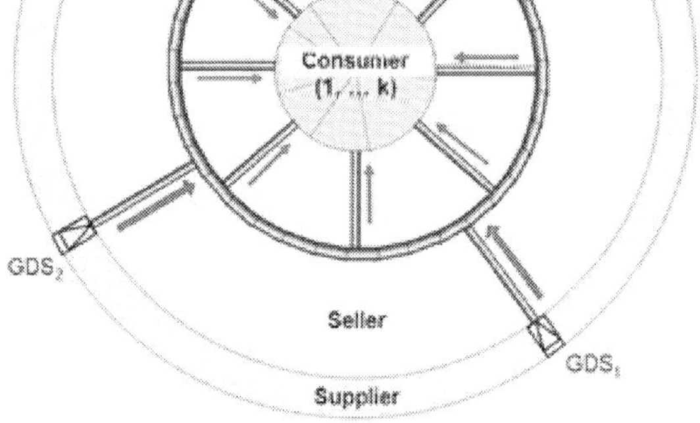

Figure 7: A diagram of the GDN under consideration

For settlement of accounts, consumers submit reports to the seller, in which they indicate estimated volumes of received gas. These reports are usually generated either by processing the consumers' field flow meter readings or by simplified calculations based on the rates formally established for the given category of consumers. Verification of data provided by the consumers consists in the comparison of their estimates with data obtained by processing the seller's flow meter readings in compliance with current guidelines.

The central difficulty in such verification is that the amount of field measurements of supplied gas that can be used as a reliable basis is rather limited in the present-day gas industry. Such a situation results in occasional discrepancies (especially during the heating season) in analyzing the volume of natural gas supplied to the consumers. The total discrepancy over a given time period is determined as a difference between two estimates of the gas volume. The first estimate represents the total gas volume actually received during the time period in question as reported by all consumers, and the second estimate, the total volume of natural gas delivered by the supplier to the seller less the gas volume accumulated in the GDN.

One of the most promising ways to resolve the above problem is to use the CFD-simulator. For this purpose, the following problem setup can be used for numerical monitoring of gas distribution discrepancy using CFD-simulator.

Input data: layout chart of the GDN; sensor locations in the GDN, where gas parameters are measured; given time interval of GDN operation; results of field measurements of gas parameters in the GDN in the given time interval; actual (or nameplate) errors of instruments used to measure gas parameters; data on received gas volumes as reported by each consumer for the given time interval.

Target data: (1) physically based gas flow parameters in the GDN in the given time interval having a minimum discrepancy compared to respective field measurement data at identification points and providing the closest possible agreement between calculated flow rate values at the outlet of each associated CGP and corresponding reported values (further as the text goes, this mode will be called "the identified gas flow"); (2) associated CGPs with underreported gas volumes as against the identified gas flow; (3) calculated estimates of discrepancies between gas volumes delivered in the given time interval through each associated CGP as an arithmetic difference between the calculated gas volume corresponding to the identified gas flow and the reported value; (4) calculated estimates of discrepancies between gas volumes delivered in the given time interval through each inlet GDS as an arithmetic difference between the calculated gas volume corresponding to the identified gas flow and the reported value.

Correct simulation of item 1 in the problem statement makes it possible to obtain credible information on physically consistent space-time distributions of flow rates, pressures and temperatures for the gas flow, which is most reasonable for the given time interval with the given field measurement data. Convergence of calculated and reported gas flow values for individual consumers increases the level of objectivity of numerical analysis, as it seeks to maintain the highest possible trust in the data on received gas volumes reported by the consumers.

It follows from the above problem statement that numerical monitoring of gas distribution discrepancy under items 2–4 in the list of target values in essence consists in performing straightforward arithmetic operations with output data of item 1. Therefore, special attention below will be paid to the algorithm of this calculation. This algorithm was proposed by V. Seleznev in 2008. In the first approximation, we consider the process of gas flow through the GDN to be steady-state.

In order to calculate non-isothermal steady-state gas flow parameters in the GDN under consideration, the following boundary conditions of "Type I" need to be specified: pressure, temperature and composition are defined at the outlet of each inlet GDS; mass flow rate and gas temperature are defined at the outlet of each associated CGP.

Using the CFD-simulator with the given boundary conditions and fixed GDN characteristics, one can unambiguously determine physically based spatial distributions of calculated estimates of steady-state GDN operation parameters (Seleznev et al., 2007). Spatial distributions of parameters here mean their distributions along the pipelines.

A diagram of identification locations is generated on the given layout of sensor locations in the GDN. The preferred location of each identification point should correspond to the key requirement: a considerable change in the fluid dynamics conditions of GDN operation should be accompanied by considerable changes in the gas parameters actually measured at this point. The distribution of identification points over the GDN diagram should be as uniform as possible. An identification point can be located both inside the GDN and at its boundaries. At each identification point, different combinations of major gas flow parameters can be measured. These combinations can be varied for every identification point.

The process of finding the identified gas flow comes to the statement and solution of the problem of conditional optimization:

$$\left\| \vec{\mathbf{f}}_{\text{calc}}\left(\vec{\mathbf{X}}\right) - \vec{\mathbf{f}}^{\,\text{const}}_{\text{meas}} \right\|_L \to \min_{\vec{\mathbf{X}}\in\Omega\subset R^n}$$

where ‖…‖L is the vector norm, the type of which is determined by the value of the parameter$L$(L=012)(see below); **f**⃗ calc(**X**⃗) $f$calc:Rn→Rm is the vector-function of calculated estimates of controlled transported gas variables at the identification points in the m-dimensional Euclidean space Rm (these calculated estimates are obtained using the CFD-simulator); **f**⃗ constmeas∈Rm is a given vector of measured values of controlled transported gas variables at the identification points; $m$is the number of given identification points in the GDN diagram; **X**⃗ ∈Ω⊂Rn is the vector of independent controlled variables in the n-dimensional Euclidean space Rn;

$$\vec{\mathbf{X}} \in \Omega = \left\{ \vec{\mathbf{X}} \in R^n : \vec{\mathbf{a}} \le \vec{\mathbf{X}} \le \vec{\mathbf{b}} \left\| \vec{\mathbf{q}}^{\,\text{GDS}}_{\text{calc}}\left(\vec{\mathbf{X}}\right) - \vec{\mathbf{q}}^{\,\text{const}}_{\text{meas\_GDS}} \right\|_0 \le t^{\text{GDS}}_{\text{flow\_rate}} \right\}$$

a⃗ ∈$Rn$and**b**⃗ ∈Rn are correctly defined vectors setting limits in simple constraints for the range of admissible variation of the vector of independent controlled variables (see below); $n$is the number of independent controlled variables (see below); ‖…‖0 is the cubic vector norm (e.g.,‖‖**Y**⃗ ‖‖0=max1≤i≤n | yi | **Y**⃗ ∈Rn **q**⃗ GDScalc(**X**⃗) qGDScalc:Rn→Rl is the vector function of calculated estimates of mass flow rates through inlet GDSs in the *l*-dimensional Euclidean space *Rl* (these calculated estimates are obtained using the CFD-simulator); **q**⃗ constmeas_GDS∈*Rl* is a given vector of measured mass flow rates at GDS outlets; *l*is the number of GDSs; *t*GDSflow_ rate=*const* is a given upper estimate of actual (nameplate) absolute error of flow meters installed at the inlet GDSs. The constraint in the form of a one-sided weak inequality in (50) formalizes the assumption that the probability of gas underdelivery by the supplier is small.

Components *xi* of the vector of independent controlled variables here mean some boundary conditions of "Type I" specified in the simulations of steady-state fluid dynamics conditions using the CFD-simulator. As practice shows, problem (49, 50) can be solved successfully, if as components of the vector of independent variables one uses an integrated set of mass flow rates at outlet boundaries of associated CGPs (*xii*=1*k*) and pressures at outlet GDSs(*xii*=*k*+1*n* *n*=*k*+*l*), where *k* is the number of associated CGPs

Components (*aibii*=1*k*) (see (50)) establish the ranges for controlled variables, the size of which is largely attributed to the degree of the seller's actual trust in a certain consumer. The following conditions should be necessarily observed:

$$a_i + t^{\text{cons}}_{\text{flow\_rate}} \quad \left[q^{const}_{cons}\right]_i b_i - t^{\text{cons}}_{\text{flow\_rate}} \quad i = \overline{1k} a_i + t^{\text{cons}}_{\text{flow\_rate}} \quad \left[x_0\right]_i b_i - t^{\text{cons}}_{\text{flow\_rate}} \quad i = \overline{1k}$$

$$\sum_{i=1}^{k} \left[x_0\right]_i = \sum_{i=1}^{k} \left\{ \left[q^{\text{const}}_{\text{cons}}\right]_i + \left[\Delta q^{\text{const}}_{\text{sell}}\right]_i \right\} = \sum_{j=1}^{l} \left[q^{\text{const}}_{\text{meas\_GDS}}\right]_j$$

where **q**⃗ constcons∈*Rk* is a given vector of mass flow rates at outlet boundaries of associated CGPs;*t*consflow_rate=*const* is a given upper estimate of the actual (nameplate) absolute error of flow meters installed at outlet boundaries of associated CGPs;**X**⃗0∈*Rn* is the starting point of the conditional optimization problem; **Δq**⃗ constsell∈*Rk* is the increment vector for reported mass flow rates at outlet boundaries of associated CGPs, which is chosen by the gas seller depending on the degree of trust in a certain consumer. Fulfillment of conditions

(51) is a guaranty for the consumers that the discrepancy analysis will necessarily account for their reported values of received gas volumes. Constraints (52) serve to implement quasi-steady-state operating conditions of the pipeline network of interest from the very starting point of the conditional optimization problem. The values of remaining components (*aibii*=*k*+1*n*) are generally defined in accordance with conditions:

$$a_i + t^{\mathrm{GDS}}_{\mathrm{pressure}}\ [x_0]_i\ b_i - t^{\mathrm{GDS}}_{\mathrm{pressure}}\ i = \overline{k+1n}[x_0]_i = \left[p^{\mathrm{const}}_{\mathrm{meas\_GDS}}\right]_{i-k}\ i = \overline{k+1n}$$

where $\vec{\mathbf{p}}$ constmeas_GDS∈*Rl* is a given vector of measured pressures in the GDS; *t*GDSpressure is a given upper estimate of the actual (nameplate) absolute error of pressure gauges installed in the GDS. As a result of fulfillment of conditions (51) and (53), the starting point of optimization problem (49, 50) will be the inner point with respect to simple constraints on controlled variables, which by far extends the range of methods that can be used for conditional minimization.

Thus, based on (51–53):

$$a_i \left\{ \min\left([x_0]_i - t^{\mathrm{cons}}_{\mathrm{flow\_rate}}\ \left[q^{\mathrm{const}}_{\mathrm{cons}}\right]_i - t^{\mathrm{cons}}_{\mathrm{flow\_rate}} \right) i = \overline{1k} \left[p^{\mathrm{const}}_{\mathrm{meas\_GDS}}\right]_{i-k} - t^{\mathrm{GDS}}_{\mathrm{pressure}}\ i = \overline{k+1n} \right\}$$
$$b_i \left\{ \max\left([x_0]_i + t^{\mathrm{cons}}_{\mathrm{flow\_rate}}\ \left[q^{\mathrm{const}}_{\mathrm{cons}}\right]_i + t^{\mathrm{cons}}_{\mathrm{flow\_rate}} \right) i = \overline{1k} \left[p^{\mathrm{const}}_{\mathrm{meas\_GDS}}\right]_{i-k} + t^{\mathrm{GDS}}_{\mathrm{pressure}}\ i = \overline{k+1n} \right\}$$

Problem (49–54) can take different forms depending on the type of the vector norm chosen in (49). For example, if we choose the cubic vector norm (*L*=0), we come to a discrete minimax problem with constraints in the form of one-sided weak inequalities and simple constraints on independent controlled variables:

$$\max_{1 \le i \le m} \left| \left[f_{\mathrm{calc}}\left(\vec{\mathbf{X}}\right) \right]_i - \left[f^{\mathrm{const}}_{\mathrm{meas}} \right]_i \right| \to \min_{\vec{\mathbf{X}} \in \Omega \subset R^n}$$

Solution to (55) provides so-called uniform agreement between calculated estimates of gas flow parameters and their measured values. Choosing the octahedron vector norm (L=1) transforms initial problem (49–54) into a general non-linear programming problem represented in the following way:

$$\begin{cases} \sum_{i=1}^{m} \left| \left[f_{\text{calc}}\left(\vec{\mathbf{X}}\right) \right]_i - \left[f_{изм}^{const} \right]_i \right| \to \min \\ \vec{\mathbf{X}} \in \Omega^* = \left\{ \vec{\mathbf{X}} \in R^n : \vec{\mathbf{a}} \leq \vec{\mathbf{X}} \leq \vec{\mathbf{b}} \left| \left[q_{\text{calc}}^{\text{GDS}}\left(\vec{\mathbf{X}}\right) \right]_j - \left[q_{\text{meas\_GDS}}^{\text{const}} \right]_j \right| - t_{\text{flow\_rate}}^{\text{GDS}} \leq 0 j = \overline{1l} \right\} \end{cases}$$

Choosing the Euclidean vector norm (L=2) in (49) results in the statement of a new conditional optimization problem, which is almost equivalent to (50-54, 56):

$$\sqrt{\sum_{i=1}^{m} \left(\left[f_{\text{calc}}\left(\vec{\mathbf{X}}\right) \right]_i - \left[f_{\text{meas}}^{\text{const}} \right]_i \right)^2} \to \min_{\vec{\mathbf{X}} \in \Omega^*}$$

Solution of (50-54, 57) provides root-mean-square agreement between calculated estimates of gas flow parameters and their measured values. It should be stressed here that statement (50-55) is stricter than (50-54, 57).

Problems (50-55), (50-54, 56) and (50-54, 57) can be solved numerically using the method of modified Lagrange functions (Vasilyev, 2002), which is quite suitable for this purpose. Note that in practice the time of numerical solution of (50-54, 57) in most cases is much shorter than the time of numerical solution of problems (50–55) or (50-54, 56).

To choose a certain type of the target function in problem (49, 50), a series of numerical experiments were conducted and more than a hundred applied tasks were simulated. The best (in terms of the accuracy/runtime ratio) results in simulating the identification problem (49, 50) were obtained using target function (57).

Based on the above considerations, in order to provide efficiency and improved accuracy of industrial applications, it is reasonable to propose the following algorithm for finding the identified gas flow in the GDN at the initial stage:

Define the starting point $\mathbf{X}^{\rightarrow}$ $0 \in Rn$ in accordance with conditions (52) and (53). Define the vectors $\mathbf{a}^{\rightarrow}$ *and*$\mathbf{b}^{\rightarrow}$ in simple constraints according to (54).

Solve optimization problem (50-54, 57). Results of its numerical solution become input data in searching for the conditional minimum at Step 4.

Analyze correctness of solution results from Step 2. The correctness criterion in this case is the condition of necessary fulfillment of all constraints in problem (50-54, 57). If this criterion is satisfied, proceed to Step 4. If not, extend the variation range of independent variables with subsequent transition to Step 2, i.e. ($\{\mathbf{a}^{\rightarrow}\ \mathbf{b}^{\rightarrow}\} \Rightarrow \{\mathbf{a}^{\rightarrow} * \mathbf{b}^{\rightarrow} *\}$). Usually, the range extension algorithm used here is heuristic and bases on the experience gained in the course of actual simulations.

Find numerical solution to problem (50–55) from the starting point, obtained at Step 2, by the method of modified Lagrange functions. Execution of Step 4 makes it possible to reduce or completely eliminate individual local peaks in discrepancy between calculated estimates and measured values, which may appear at Step 2.

Analyze correctness of the results obtained at Step 4, i.e. check the necessary fulfillment of all constraints in problem (50–55). If the correctness criterion is not fulfilled, solution of Step 3 is assumed to be the target solution.

The vector of controlled variables corresponding to the optimal solution at Step 5, is designated as $\mathbf{X}^{\rightarrow}$ init, with $\mathbf{X}^{\rightarrow}$ init$\in \subset Rn$. The found fluid dynamics conditions of GDN operation is taken as the primary fluid dynamics mode. Its calculated parameters have uniform (i.e. strictest) agreement with respective measured values.

At the final stage of identification, the primary fluid dynamics mode is corrected within the available measured information, in order to minimize possible discrepancies between calculated and reported estimates of gas volumes transmitted through each associated CGP in the given time interval. This stage is legal by nature, because given the limited amount of measured data the gas seller has no right to accuse the consumer a priori of deliberate misrepresentation of reported received gas volumes. This stage consists in the solution of the general nonlinear programming problem:

$$\begin{cases} \left\| \vec{\mathbf{q}}^{\,\mathrm{cons}}_{\mathrm{calc}}\left(\vec{\mathbf{X}}\right) - \vec{\mathbf{q}}^{\,\mathrm{const}}_{\mathrm{cons}} \right\|_2 \to \min \\ \vec{\mathbf{X}} \in \Theta = \left\{ \vec{\mathbf{X}} \in R^n : a_i^{**} \le x_i \le b_i^{**}\ i = \overline{1k} \right. \\ \left[p^{\mathrm{const}}_{\mathrm{init\_GDS}} \right]_i - t^{\mathrm{GDS}}_{\mathrm{pressure}} \le x_i \le \left[p^{\mathrm{const}}_{\mathrm{init\_GDS}} \right]_i + t^{\mathrm{GDS}}_{\mathrm{pressure}}\ i = \overline{k+1n} \\ \left| \left[f^{\mathrm{ident}}_{\mathrm{calc}}\left(\vec{\mathbf{X}}\right) \right]_s - \left[f^{\mathrm{const}}_{\mathrm{init\_ident}} \right]_s \right| - t^{\mathrm{ident}}_{\mathrm{pressure}} \le 0 s = \overline{1h} \\ \left| \left[q^{\mathrm{GDS}}_{\mathrm{calc}}\left(\vec{\mathbf{X}}\right) \right]_j - \left[q^{\mathrm{const}}_{\mathrm{meas\_GDS}} \right]_j \right| - t^{\mathrm{GDS}}_{\mathrm{flow\_rate}} \le 0 j = \overline{1l} \left. \right\} \end{cases}$$

where **q**⃗ conscalc(**X**⃗)⊂**X**⃗ [*q*calc(*x*)]*i*=*xii*=1*k* is the vector function of calculated mass flow rates for outlet boundaries of associated CGPs in the *k*-dimensional Euclidean space *Rk* **p**⃗ constinit_GDS∈*Rl* is a given vector of GDS pressure corresponding to the primary fluid dynamics mode at **X**⃗ init∈*Rn* **f**⃗ identcalc(**X**⃗)*f*identcalc:*Rn*→*Rh* is the vector function of calculated estimates of controlled variables at inner identification points in the *h*-dimensional Euclidean space *Rh* (these calculated estimates are obtained using the CFD-simulator); **f**⃗ constinit_ident∈*Rh* is a given vector of controlled variables at inner identification points corresponding to the primary fluid dynamics mode at **X**⃗ init∈*Rn h*is a given number of inner identification points; *t*identpressure=*const* is a given upper estimate of the actual (nameplate) absolute error of pressure gauges at inner identification points.

The first group of simple constraints on the controlled variables in (58) is partly redundant. It assures that numerical search for solutions in industrial applications is always performed in the domain of practically significant results. The second group of simple constraints and the second group of one-sided weak inequality constraints in problem (58) account for the imperfectness of corresponding existing instruments in favor of consumers. The first group of one-sided weak inequality constraints in (58) formalizes the demand for the closest possible uniform agreement between calculated estimates and reported volumes of gas received by each consumer.

Problem (58) can be solved using modified Lagrange functions (Vasilyev, 2002). As a starting point here we use $\vec{X}$ init$\in Rn$. The target result of the simulation should necessarily be correct, i.e. it should fulfill all simple constraints and inequality constraints of problem (58). Otherwise, the primary fluid dynamics mode is taken as a solution to (58).

The simulation outcome of optimization problem (58) is the final solution to the problem of finding the identification gas flow in the GDN. The target identified gas flow is completely defined by the vector $\vec{X}$ ident$\in\Theta\subset Rn+1$, corresponding to the optimal solution of problem (58), and is characterized by the fulfillment of the following conditions: (1) calculated gas flow parameters at each identification point should be as close as possible to corresponding field measurement data; (2) calculated estimates of gas volumes supplied to the GDN in a given time interval should correspond to the supplier-reported values within actual (nameplate) absolute errors of the flow meters installed in the GDS; (3) calculated estimates of gas volumes received by each consumer in the given time interval should be as uniformly close as possible to the values reported by the consumers.

CONCLUSION

The methods and approaches described in this chapter have been developed during the past fifteen years. These methods demonstrated their efficiency as applied to simulations for Gazprom, Russia, and SPP, a.s., Slovakia (Seleznev et al., 2005). These methods apply to pipeline systems for transportation of liquid and gas-liquid products. They are rather efficient in analysis and prevention of accidents (Aleshin & Seleznev, 2004).

The approach presented in this article for high-accuracy numerical analysis of operating parameters of industrial pipeline networks using CFD-simulators is based on adaptation of the full system of equations of fluid dynamics to conditions of transient, non-isothermal processes of the flow of gas mixtures in actual GTNs. The adaptation applies the rule of minimization of the number and depth of accepted simplifications and assumptions. The high accuracy of analysis of industrial pipeline networks operating parameters is understood here as the most reliable description and prediction of actual processes in a GTN, which are achievable due to the present level of development of mathematical modeling and technical monitoring methods and available computer hardware. Development and operation of CFD-simulators in solving industrial problems of improving safety, efficiency and environmental soundness of pipeline network operation can be regarded as one of the promising trends of industrial application of the state-of-the-art computational mechanics methods.

In the past few years, the proposed approach to the numerical monitoring of natural gas delivery through gas distribution networks has proved to be well-performing in industrial applications for discrepancy analysis of natural gas deliveries to large and medium-size consumers. This approach can be implemented even on standard personal computers.

Future development of the methods discussed will be largely focused on their applications in high accuracy analysis and control of multi-phase fluid transportation.

REFERENCE

1. V. V. Aleshin, V. E. Seleznev, 2004 Computation technology for safety and risk assessment of gas pipeline systems, Proceedings of the Asian International Workshop on Advanced Reliability Modeling (AIWARM'2004), 443 450 , 9-81238-871-0 Japan, August 2004, World Scientific Publishing Co. Pte. Ltd., London.
2. J. E. Dennis, R. B. Schnabel, 1988 Numerical Methods for Unconstrained Optimization and Ninlinear Equations, Prentice-Hall Inc., 0-13627-216-9 Jersy.
3. C. A. J. Fletcher, 1988 Computational Techniques for Fluid Dynamics. Fundamental and General Techniques, Springer-Verlag, 3-54018-151-2
4. V. V. Kiselev, V. E. Seleznev, O. I. Zelenskaya, 2005 Failure forecast in engineering systems by searching for the inner points of system of algebraic equalities and inequalities, Proceedings of the European Safety and Reliability Conference (ESREL 2005), 2 1773 1776 , 0-41538-343-9 City (Gdyna-Sopot- Gdansk), Poland, June 2005, Taylor & Francis Group, London.
5. V. E. Seleznev, V. V. Aleshin, G. S. Klishin, R. I. Il'kaev, 2005 Numerical Analysis of Gas Pipelines: Theory, Computer Simulation, and Applications, KomKniga, 5-48400-352-0
6. V. Seleznev, 2007 Numerical simulation of a gas pipeline network using computational fluid dynamics simulators. Journal of Zhejiang University

SCIENCE A, 8 5 755 765 , 0167-3565X (Print); ISSN 1862-1775 (Online)

7. V. Seleznev, S. Pryalov, 2007 Numerical forecasting surge in a piping of compressor shops of gas pipeline network. Journal of Zhejiang University SCIENCE A, 8 11 1770 1783 , 0167-3565X (Print); ISSN 1862-1775 (Online)
8. V. E. Seleznev, V. V. Aleshin, S. N. Pryalov, 2007 Mathematical modelling of pipeline networks and channel systems: methods, algorithms, and models, MAX Press, 978-5-31702-011-8 Moscow. (In Russian)
9. N. Tikhonov, A. A. Samarsky, 1999 Equations of Mathematical Physics, Moscow State University Publisher, 5-21104-138-0 (In Russian)
10. F. P. Vasilyev, 2002 Methods for Optimization, Factorial Press, 588688059 (In Russian)

Chapter 5

TURBULENCE, VIBRATIONS, NOISE AND FLUID INSTABILITIES. PRACTICAL APPROACH.

Carlos Gavilán Moreno[1]

[1]Cofrentes N.P.P. Iberdrola S.A.,Spain

INTRODUCTION

Colloquially speaking, turbulence in any language means disorderly, incomprehensible, and of course, unpredicTable movement. Consequently, we encounter expressions that employ the word turbulence in social and economic contexts; in aviation whenever there are abnormalities in the air, and even in psychology and the behavioral sciences in reference to turbulent conduct, or a turbulent life, in the sense of a dissolute existence. Thus has the word turbulence become associated with chaos, unpredictability, high energy, uncontrollable movement: dis sipation. All of the foregoing concepts have their source in the world of hydrodynamics, or fluid mechanics.

In fluid mechanics, turbulence refers to disturbance in a flow, which under other circumstances would be ordered, and as such would be laminar. These disturbances exert an effect on the flow itself, as well as on the elements it contains, or which are submerged in it. The process that is taking place in the flow in question is also affected. As a result, they possess beneficial properties in some fields, and harmful ones in others. For example, turbulence improves processes in which mixing, heat exchange, etc., are involved. However, it demands greater energy from pumps and fans, reduces turbine efficiency and makes noise in valves and gives rise to vibrations and instabilities in pipes, and other elements.

The study of turbulence and its related effects is a mental process; one that begins with great frustration and goes on to destroy heretofore accepted theories and assumptions, finally ending up in irremediable chaos. "I am an old man now, and when I die and go to heaven, there are two matters on which I hope for enlightenment. One is the quantum electrodynamics, and the other is the turbulent motion of fluids. An about the former I am rather optimistic" (Attributed to Horace Lamb).

Nonetheless, some progress has been made in turbulence knowledge, modelling and prediction. (Kolmogorov, 1941). This chapter will deal briefly with these advances, as well as with the effects of turbulence on practical applications. In this sense, reference will be made to noise effects and modelling, as well as to flow vibration and instabilities provoked by turbulence. (Gavilán 2008, Gavilán 2009).

TURBULENCE

Of itself, turbulence is a concept that points to unpredictability and chaos. For our purposes, we will deal with this concept as it applies to fluid mechanics. Therefore, we will be dealing with turbulent flow. Throughout what follows, the terms *turbulence* and *turbulent flow* will be understood as synonymous. Some texts treat the terms *turbulence* and *vortex* as analogous, however, this seems to be rather simplistic. For the purposes of this work, it seems best to take the concept of turbulence in its broadest sense possible. Historically, fluid mechanics has been treated in two different ways, namely, in accordance with the Euler approach, or pursuant to the Lagrange approach. The Euler Ian method is static, given that upon fixing a point, fluid variations are determined on the effect they have on this point at any given time. On the other hand, the Lagrangian method is dynamic, given that it follows the fluid. In this way, variations in the properties of the fluid in question are observed and/or calculated by following a particle at every single moment over a period of time.

The Euler Ian method is the one most employed, above all, in recent times, by means of numerical methods, such as that of finite elements, infinites, finite volumes, etc. Notwithstanding, there is great interest in the Lagrangian method or approach, given that it is one that is compatible with methods that do not use mesh or points. (Onate, E; et al. 1996)

Throughout history, there have been two currents of thought as regards the treatment of turbulence. One is the so-called deterministic approach, which consists of solving the Navies-Stokes equation, with the relevant simplifications, (Euler, Bernoulli) practically exclusively via the use of numbers. The other approach is statistical. The work of Kolmogorov stands out in this field; work which will be dealt with below, given its later influence on numerical methods and the results of same. Apart from the Eulerian or Lagrangian methods, classic turbulent fluid theory will be dealt with in Section 2.1, whereas the statistical or stochastic approach will be dealt with in Section 2.2, in clear reference to Kolmogorov's theory. (Kolmogorov, 1941).

GENERAL THEORY

This section provides a brief and concise exposition of successive fluid flow approaches designed to respond to the presence of anomalies that were later referred to as *turbulence*, and which gave rise to the concept of turbulent fluid. Furthermore, the equations given enable the visualization of the turbulence in question and its later development. A Euler Ian and deterministic focus will be followed in this section.

Working in reverse to the historical approach, the fluid flow equation formulated by Navies-Stokes in 1820 is given; firstly, because it is the most general one, and secondly, because, of itself, its solution can represent the turbulent flow equation.

$$\frac{\partial u_i}{\partial t} + \sum_{j=1}^{n} u_j \frac{\partial u_i}{\partial x_j} = v\Delta u_i - \frac{\partial p}{\partial x_i} + f_i(x,t)$$

$$\frac{\partial u_i}{\partial t} + \sum_{j=1}^{n} u_j \frac{\partial u_i}{\partial x_j} = - \frac{\partial p}{\partial x_i} + f_i(x,t)$$

$$u(x,0) = u^0(x)$$

Where *u*i stands for the velocity components at each point and at each moment in time, *v* is the viscosity, *p* is the pressure at each point and at each moment in time, and *fi* refers to the external forces at each point and at each moment in time. By annulling the viscosity and its effects, we get Euler's fluid flow equation announced in 1750.

$$\frac{\partial u_i}{\partial t} + \sum_{j=1}^{n} u_j \frac{\partial u_i}{\partial x_j} = - \frac{\partial p}{\partial x_i} + f_i(x,t)$$

$$div\vec{u} = \sum_{i=1}^{n} \frac{\partial u_i}{\partial x_i} = 0$$

$$u(x,0) = u^0(x)$$

In addition to this simplification, if we make the fluid stationary, so that:

$$\frac{\partial u}{\partial t} = 0$$

we get the first Bernoulli theorem announced in 1738.

$$\frac{D}{Dt}\left(p + \rho \cdot \phi + \rho \frac{u^2}{2}\right) = 0$$

The solving of these equations, given their simplifications and context conditions, provide us with a field of speeds and pressures for a fluid in movement. Indeed, as regards the Navies-Stokes equation, it can be solved and turbulences and instabilities determined by employing powerful numerical solution methods, such as that of the finite element.

Thus, the equations that govern fluid movement. By means of these two equations, particularly the last two (those of Euler and

Bernoulli), it was observed that, under certain conditions, the results did not correspond to reality, on account of a certain problem of disorder developing in the fluid and its flow. Only the accurate and numerical solving of the Navies-Stokes equation can exactly reproduce these phenomena. To this end, resort must be had to potent computational fluid dynamic (CFD) software. Notwithstanding, in 1883, Osborne Reynolds discovered a parameter that predicted or anticipated the chaotic and turbulent of the fluid: the Reynolds number.

$$\mathrm{Re} = \frac{\rho \cdot v_s \cdot D}{\mu} = \frac{v_s \cdot D}{v}$$

Thus was it established that the flow is stationary, and therefore laminar, for Re<2000 values, a fact which meant that the solutions given by Bernoulli and Euler were very accurate. For values of 2000<Re<4000 the system was deemed to be in transition, and therefore, not stationary. The functions that work best are those of Euler and Navies-Stokes. This turbulence undergoes three phases or states of development.

- đ Growth of coherent dimensional vortices[1]
- đ Joining of Vortices
- đ Separation of vortices and turbulent state in 3D

Finally, if completely turbulent, the fluid is non-stationary and tri-dimensional for values of Re>4000. These solutions are only possible by means of using the Navies-Stokes equations.

In conclusion, only the numerical solution of the Navies-Stokes equation provides a solution that considers turbulence. Nevertheless, there are other processes and approaches that will be developed in Section 2.3. Turbulence is, therefore, produced by the interaction of the fluid with geometry, by the loss of energy due to viscosity, by density variations caused by temperature, or other factors, such as changes in speed, or all of these at once. Consequently, the turbulent flow is unpredicTable and chaotic in the sense that it depends on a host of small variations in the initial conditions and these disturbances are amplified in such a way that it becomes possible to predict them in space and time. Another of its features is its great capacity for mixing and, lastly, that it affects at various scales and wavelengths. It could be said that together, fluid, structure and context conditions, constitute a non-linear, non-stationary dynamic system. Its most noteworthy characteristic is its sensitivity to the initial conditions and its self-similarity, which will serve as a starting point to develop Kolmogorov's theory.

In the 19th and 20th centuries, several researchers devoted great efforts to studying turbulence under certain, extremely particular conditions. Examples of such turbulence or instabilities are to those of Von Karman, Kevin-Helmets, Raleigh-Bernard, and so on.

KOLMOGOROV´S THEORY

Kolmogorov's theory cannot be dealt with without first mentioning the spectral analysis of turbulence, or the application of Fourier's analysis to the study of turbulence. Fourier's theory decomposes the fluctuations into sinusoidal components and studies the distribution of the turbulent energy along several wavelengths. In this way it becomes possible to get several scales of turbulence and their evolution in time. This technique works and produces accepTable results when turbulence is homogenous. Under this condition, accurate equations can be determined for the speed spectrum and for the transferring of energy between difference scales of turbulence, as well as the dissipation of turbulent energy due to viscosity. The simplest development assumes that there is no average speed gradient, in such a way that the turbulence interacts with itself, with the energy dropping by itself. Neither energy sources nor sinks are taken into account.

To start, we assume that the speed of a particle in the fluid can be decomposed into an average speed plus a fluctuation component.

$$U_i = \overline{U}_i + u_i$$

R ij is defined as the speed correlation function by the expression:

$$R_{ij}(x, x', t) = \overline{u(x,t)_i u(x',t)_j}$$

Given that we assume homogenous turbulence, *R* ij is only a function of the distance *r*, which is the distance between *x* and *x'*:

$$R_{ij}(r, t) = \overline{u(x,t)_i u(x',t)_j}$$

Therefore, we assume that the value of the correlation function tends to zero when the radius tends to the infinite. We now define a spectral function *Φ* ij as the Fourier transform of the correlation function in 3D:

$$\varnothing_{ij}(k,t) = \frac{1}{(2\pi)^3}\int R_{ij}(r,t)\cdot e^{-ikr}\cdot d^3k$$

This spectral function, which will form the spectral matrix, depends on time and on the wave number, which is a vector. The turbulent kinetic energy can be expressed as:

$$\frac{1}{2}\overline{u_i u_i} = \frac{1}{2}R_{ij}(0,t) = \frac{1}{2}\int \varnothing_{ii}(k,t)d^3k$$

The magnitude of the vector of the wave number *k=[k]* is the inverse of the length, thus the length is *k* -1. Therefore, if *L* is a characteristic scale of turbulence, the *L* -1 values of the wave number represents the higher scales of turbulence, where all the turbulent kinetic energy is contained. The spectral function, *Φ* ij, describes the energy distribution in relation to wave numbers. It has a maximum value of *[k]*=*L* -1, which drops to zero when *[k]* tends to the infinite. The evolution of kinetic energy is defined as:

$$E = 2\pi k^2 \varnothing_{ii}$$

As far as the evolution of the *k* energy value is concerned, this can be seen in the following Figure (Figure 1).

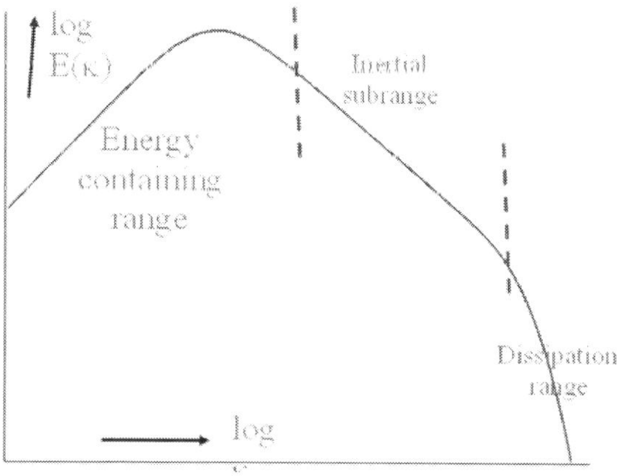

Figure 1. Energy ranges.

This is where the term, or value, η comes in, referring to Kolmogorov's scale, which determines the scale point at which energy dissipation by viscosity is most significant. We will use the following equation for Kolmogorov's theory:

$$\frac{\partial E}{\partial t} = T - 2 \cdot v \cdot k^2 E$$

That is to say, the energy and its evolution in time will be a function of that which is transferred and that which is dissipated. Integrating it from k=0 to a greater, or finite, as well as limited, value gives us:

$$\frac{d}{dt} \int_0^k E(k' \cdot t)\, dk' = \aleph(k,t) - 2v \int_0^k k'^2 E(k',t)\, dk'$$

Where

$$\aleph(k,t) = -\int_0^k T(k',t)\, dk' = \int_k^\infty T(k',t)\, dk'$$

This can be interpreted as follows:

"Energy changes in wave numbers below *k* on account of the interchange with larger wavelengths through the energy spectral fluid and the dissipation caused by viscosity forces on its own wavelength"

Kolmogorov holds that turbulence properties in the inertial range, above η, only depend on the energy spectral fluid, in such a way that:

$$\aleph = \varepsilon$$

The idea behind this is that the transfer cascade evolves towards lower scales, leaving the spectral fluid as a fixed parameter. Consequently, the energy equation now stands at:

$$E(k,t) = C \cdot \varepsilon^{-\frac{2}{3}} \cdot k^{-\frac{5}{3}}$$

Where *C* is a Kolmogorov constant, considered as a universal, the value of which is 1.5 when *L*-1<<*k*<<η -1 and the *Re* (Reynolds number) is high. Thus, beyond the inertial scale wave numbers, we have the dissipation range numbers, as shown in Figure 2.

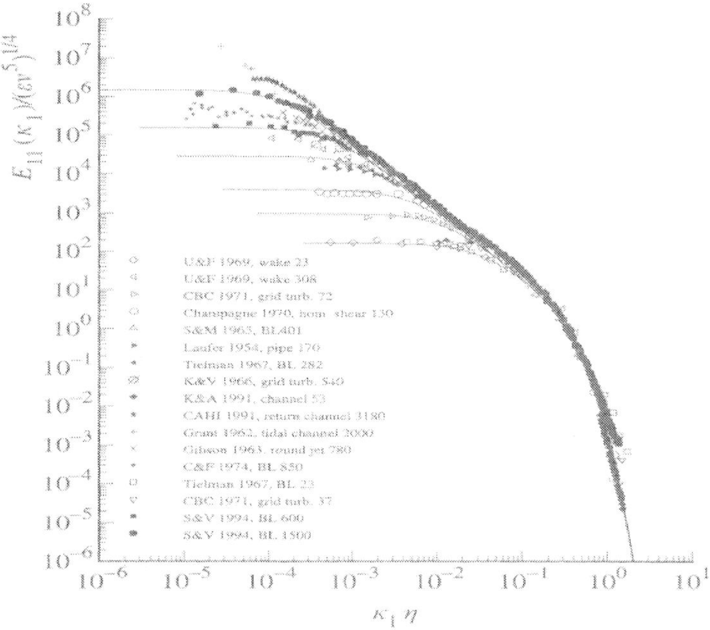

Figure 2. Kolmogorov scales and data from several tests.

This range is only determined by the η parameter, which only has a dependence parameter, as shown by the equation (21).

$$\eta = (\underline{\upsilon})^{\frac{1}{4}}$$

This Kolmogorov derived value gives the size of the smallest turbulence dissipation scale. The dissipation range energy expression now becomes:

$$E(k,t) = \varepsilon^{\frac{1}{4}} \upsilon^{\frac{5}{4}} \cdot F(k\eta)$$

Where *F ( )* is also a universal function.

We will define the Kolmogorov time and speed scale factor with the ε y υ parameters, which determine the behavior of the flow in the dissipation range:

$$t_\eta = \left(\frac{\upsilon}{\varepsilon}\right)^{\frac{1}{2}}$$

$$u_{\eta} = (\upsilon \cdot \varepsilon)^{\frac{1}{4}}$$

Thus, the Kolmogorov theory completes the spectral analysis. The theory postulates high Reynolds number values. The small turbulence scales are assumed that serve to balance and be controlled by the average energy flow, which is generated in the inertial scale and which equals the dissipation rate. Furthermore, Kolmogorov's theory universally predicts speed properties and their differences for small separations, as well as their correlations and spectre, only depending on the υ y ε parameters. Kolmogorov also marked the boundary between the transferred or contained energy range (inertial range) and the dissipative structures by way of the following expression:

$$\frac{L}{\eta} = f(\mathrm{Re}^{3/4})$$

Therefore, given a Reynolds number, the lower scales are not sensitive to the turbulent flow in which they find themselves. Nevertheless, the lower scales are intimately related to the flow, with their properties varying substantially depending on the specific flow. These concepts will form the basis for future methods of modeling and solving fluid movement problems, such as *k*-ε and Large Eddy Simulation (LES) models.

SIMULATION OF TURBULENT FLUIDS

As is well known, navies, L.H.H. and Stokes, G.G derived fluid movement equations over 150 years ago. This equation, along with that of continuity, provides an answer to any fluid movement problem.

The solution and determination of the speed field is, therefore, nothing more than discretizing the domain and the differential equations and applying context conditions in order to repeatedly solve the system formed until achieving convergence. This is the so-called Direct Numerical Solution (DNS). Nonetheless, this simple method is only useful in simple geometries, given that otherwise, the time calculation would be so big as to make any simulation unfeasible. This defect arises noticeably when we design fluid-structure (FIS) models. Consequently, other models need to be found, which are less costly computationally speaking.

The most common solution to the high number of elements required by the former simulation method, DNS, is to use weighted, or weighting, techniques. In this way, modeling is done on a small scale expecting that the solution will respond to the flow as a whole. This is the idea that underlies the so-called Reynolds Averaged Navier-Stokes (RANS). It must be assumed that the speed of a turbulent flow can be described as follows:

$$\vec{u} = \overline{\vec{u}} + \vec{u}'$$

That is to say, as the sum of an average speed plus a fluctuating component on this average speed. The following is an example of the application of this assumption. It refers to the 2D modeling of a turbulent flow on a flat plate, the navies Stokes (NS) equation for which is given below:

$$\rho\left(\frac{\partial u_x}{\partial t} + u_x \frac{\partial u_x}{\partial x} + u_x \frac{\partial u_x}{\partial y}\right) = \mu \frac{\partial^2 u_x}{\partial y^2}$$

Replacing

$$\rho\left(\overline{u_x} \frac{\partial \overline{u_x}}{\partial x} + u_x \frac{\partial \overline{u_x}}{\partial y}\right) = \mu \frac{\partial^2 \overline{u_x}}{\partial y^2} - \rho \frac{\partial}{\partial y} \overline{u'_x u'_y}$$

The difficulty of this expression and of all those assuming equation (28) is that an analytical expression of the term is required:

$$\rho \frac{\partial}{\partial y} \overline{u'_x u'_y}$$

Known as Reynolds Stress. In order to avoid this difficulty, the kinetic energy (k) and the dissipation rate (ε) are modeled. This gives rise to a well-known model: the k- ε model.

The simplification consists of the following:

$$\rho \cdot \overline{u'_x u'_y} = \mu_{turb} \frac{\partial \overline{u_x}}{\partial y}$$

Where, moreover,

$$\mu_{turb} = \frac{C_k \cdot k^2}{\varepsilon}$$

by replacing all of this, in the proposed movement equation, the result is as follows:

$$\left(\overline{u_x}\,\frac{\partial \overline{u_x}}{\partial x}+\overline{u_y}\,\frac{\partial \overline{u_x}}{\partial y}\right)=v\,\frac{\partial^2 \overline{u_x}}{\partial y^2}-\frac{C_k k^2}{\varepsilon}\,\frac{\partial \overline{u_x}}{\partial y}$$

This k- ε model has a big advantage over the DNS model in that it takes less time to compute. This model provides accepTable solutions when flow fluctuations are not very important. When the flow motor, or the problem, to be studied has to do with pressure fluctuation, that is to say, external flow Flow Induced Vibrations (FIV), the RANS model is unviable, given that it does not provide quality solutions. There are several RANS model variations, the main features of which are shown in Table 1.

Table 1.RANS sub models, strengths and weaknesses.

Method	Strengths	Weaknesses
Spalart-Allmaras	A one-equation model, which provides less computational effort than most other models. Producet for external flow over aitfoils but is increasing in popularity for turbo machinery applications. Performs well for attached wall-bounded flows with weakly complex boundary layers	Weak for adverse pressure gradients that produce boundary layer separation. Since it is relatively new, it has a lack of submodels available.
RNG k-ε	Possesses many of the same characteristics as the standard k- ε, but uses mathematical group theory to detrmine the previously empirical constans. It preforms better for moderately complex flows like jet impingement	Subject to limitations due to isotropic eddy viscosity assumptions

Reynolds Stress Model (RSM)	Highly rooted in the physics by solving a transport equation for each Reynolds stress.	Requires much more computational effort than any other technique.

Lastly, there is a model that lies halfway between the RANS and DNS. The DNS model has been seen to represent the real circumstance and disturbances very well. On the other hand, though the RANS model is extremely comforTable computationally speaking, it does not represent turbulence or disturbances very well. The so-called Large Eddy Simulation (LES) model provides instantaneous solutions, as does the DNS model, while at the same time containing models and simplifications such as those in the RANS model. The LES model solves the Navies-Stokes equation as does the DNS model, but the equations are spatially filtered, or refined. This filtering or refining of the equations means that the flow is determined at a characteristic scale, and is modeled afterwards at lower scales. It is the application of Kolmogorov's theory to non-stochastic numerical models. This succession of modeling and refining is due to the fact that big eddies behave anisotropic ally and, therefore, must be calculated, whereas smaller eddies behave isotropic ally and can, as such, be treated statistically. Thus, the mesh is such that the majority of energy is contained in big eddies and calculated directly, the rest is assigned in a weighted manner to flow as a whole.

The bigger the Reynolds number, the bigger the cost in this LES model. Mathematically, the model is based on the *Φ* flow being formed by two superimposed flows:

$$\phi = \bar{\phi} + \phi'$$

The mesh flow, or the grand scale flow, □‾

The sub-mesh flow or the small scale flow □'

The grand scale (GS) flow is calculated as follows:\

$$\overline{\phi(x)} = \int \phi(x')G(x, x')dx'$$

Where

$$G(x, x') = \left\{ \begin{array}{c} \frac{1}{v} \sqrt{x \epsilon \Delta v} \\ 0 resto \end{array} \right\}$$

Then replacing

$$\overline{\phi(x)} = \frac{1}{v} \int \phi(x') dx'$$

By applying these assumptions to the flow movement equation

$$\frac{\partial \overline{u_i}}{\partial t} + \frac{\partial}{\partial x} \left(\overline{u_i u_j} \right) = - \frac{1}{\rho} \frac{\partial}{\partial x} \bar{p} + v \frac{\partial^2 \overline{u_i}}{\partial x_j^2} + \tau_{ij}$$

$$\tau_{ij} = \overline{u_i u_j} - \overline{u_i} \cdot \overline{u_j}$$

the latter term is modeled on the sub-mesh, or on the SGS sub-model and the accuracy of the model rests on the idea that lower speeds than the mesh are homogenous and, therefore, can be modeled with great accuracy. The aim of the LES method is to solve the majority of the flow and to model only a small part of it. Thus, the LES method strikes a balance between mesh size and accuracy. Figure 3, in which the solution for a turbulent flow in a pipe compares the RANS method with the LES one, provides an explanatory example.

These are practically the general models that are used in flow simulation and fluid movement. Notwithstanding, there are variations of these basic methods depending on the application in question.

Figure3. Solutions of the same problem with a RANS model (on the left) and LES model (on the right).

Quite clearly, discretization's can be Euler Ian, or Lagrangian, in such a way as to respond to both forms of tackling problems concerning fluid dynamics.

General effects of the turbulence. (FIC)

Having introduced turbulence, the turbulent flow and its equations and simulations, the question arises as to what can turbulence do? What are its possible effects? To answer these questions, first off it is necessary to decide whether the effects in question are general or local, that is to say, whether they are provoked by the turbulent flow or, on the contrary, are caused by turbulence or instabilities. The latter will be dealt with in Sections 4, 5 and 6, while the general effects are considered below.

Numerous studies have been carried out highlighting the benefits of turbulence with respect to miscibility, diffusion and heat exchange, though it can also have harmful effects. Here we are going to deal with the phenomenon of spontaneous cavitation, and its effects, in turbulent flow. Specifically, our aim is to study the effect of the change of speed caused by vortices and pressure due to load loss on account of turbulence in high energy turbulent fluids near saturation point, such as, for example, feed water in power stations. Firstly, we are going to define a non-dimensional parameter, which is called the cavitation number.

$$Ca = \frac{Pa - Pv}{\frac{1}{2}\rho v^2}$$

Where:

P a is the local or system pressure;

P v is the vaporization pressure at system temperature;

P is the fluid density at system temperature;

V is the fluid speed.

Cavitation will occur when the parameter value is low. There is a limit value that corresponds to a determined speed, for each temperature and pressure, below which cavitation is ensured.

Speed can be defined in a turbulent flow as follows:

$$u_T(t) = u_m + \delta u(t)$$

That is to say, an average value and a fluctuation or oscillation component. Moreover, turbulence intensity is defined as:

$$u'^2(t) = \frac{1}{T}\int_0^T u_T(t)^2 dt$$

Replacing the equation 40 in the equation 39 gives:

$$Ca = \frac{Pa - Pv}{\frac{1}{2}\rho\left(u_m^2(t) + 2\cdot u_m(t)\cdot\delta u(t) + \delta u^2(t)\right)}$$

If we assume that $\delta u(t)$ is proportional to the standard deviation of the measured values in the speed system, and that in turn, this standard deviation is a function of the average speed, (Gavilán, C.J. 2008) we find that:

$\delta u(t)$ is zero for low speeds, therefore, the Y expression is accurate and can be applied to all fluid points. Moreover, the Bernoulli and Euler theorem is applicable, therefore, there will be no spontaneous cavitation, unless the average speed changes.

For high speeds, $\delta\ u(t) >> 0$, there may be some speed values at which *Ca* is lower than the average value, and therefore, spontaneous cavitation may occur.

This effect will be called Fluid Induced Cavitation (FIC), given that, although cavitation is associated with turbulence, in turbomachines or rotary pumps, little attention is paid to spontaneous cavitation due to the effect of turbulence on fluid systems. This effect is not widely known, though it is most definitely of great importance when dealing with fluids that are working very near the vaporization limit, or to put it more clearly, when the system pressure is very near the fluid vapour pressure at working temperature. This is particularly important for thermal power plants, or energy producing stations. In such facilities, the liquid is heated before entering the vaporization element; if there are elements that make the pressure fall, such as elbows, T's, filters, etc. in the conditions prior to vaporization, the numerator of equation 39 drops, reducing the *Ca* value, thus representing a high risk situation. In the same way, if we have elements in which the speed rises, such as jet pumps, venturi tubes, and other restrictions, the denominator increases, in such a way that the *Ca* value falls, thus increasing the risk of cavitation.

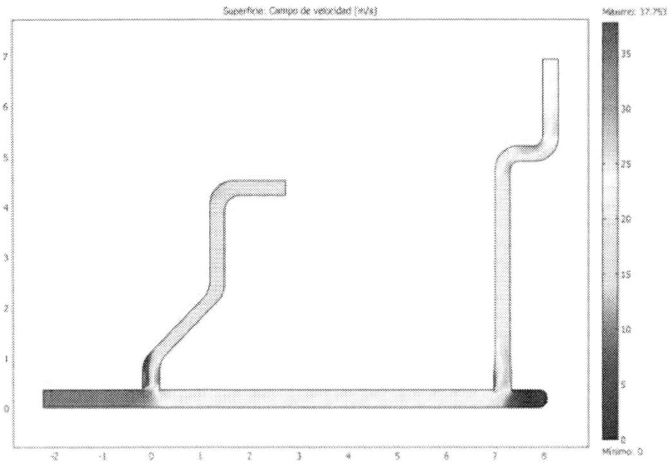

Figure 4. Flow in pipe system.

Cavitation, which elsewhere and in other fora is referred to as *flashing*, is a vibration and noise - FIV and FIN - and may even manage to change the flow pattern, and therefore, cause FII flow instabilities. Other effects also need to be taken into account, such as the presence of erosion-corrosion, in what is, in principle, a single-phase fluid. Furthermore, if there are sTable cavitation conditions, and a real and permanent void ratio is established, pump cavitation is ensured, even when this should not occur according to the calculations.

There are other reasons for possible cavitation, such as variations in the density and vapor pressure parameters on account of changes in the pressure and temperature conditions.

Let us have a look at a real example. Figure 4 shows the case of feed water speed in a pipe to an electricity-generating installation boiler. If we calculate the *Ca* parameter between two points, A and B, already having the average, or tendency, speed and the real one, we see that their *Ca* values are different. After the *Ca*, *v* and *p* values have been calculated in several situations, if we extrapolate them, there is such a turbulent speed value that it provokes spontaneous cavitation (Figure 5). Therefore, there is a speed limit for each flow at which spontaneous cavitation occurs, which coincides with the asymptotic value.

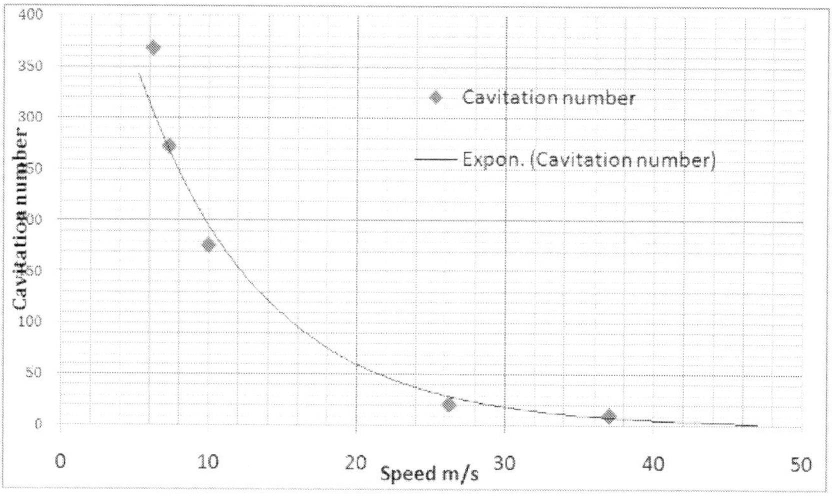

Figure 5. Cavitation number versus flow speed.

The cavitation effect is not completely harmful, given that occasionally it is provoked by means of the speed of the fluid around a vehicle in such a way that movement resistance is very low and energy is saved in the movement, the fluid entry length is increased, or the energy is increased on knocking against another object in the fluid. This is the case of the supercavitating torpedoes used by the Russian navy.

FLUID INDUCED VIBRATION. (FIV)

There are lots of books and articles that have dealt with this subject. Strictly speaking, the term was only coined in the 1980's. In truth, this case study is a particular case of fluid-structure interaction. Interest in this subject lies in the fact that the source of vibration is dissipated energy caused by turbulence, or in other cases, by eddies that produce oscillating lift forces that impregnate objects immersed in the fluid with a vibratory movement.

There are two basic FIV mechanisms:

1. a self-induced vibrating mechanism
2. a forced vibration mechanism.

Mathematically speaking, this is true. Nevertheless, the subject is somewhat wider than this might suggest, as can be seen from the following classification, which provides us with a more complete view. (Figure 6).

As far as we are concerned, two of the most important effects, or examples, as regards nuclear plants of the Boiling Water Reactor (BWR) type, are those that reflect the influence of FIV on reactor internals. That is to say, on single-phase fluids in a non-stationary operating system with turbulent flow, and in the case of a bi-phase flow, those that are subject to FIV on account of the vibrations caused by the phase change, which may even go as far as to cause thermo-hydraulic instabilities.

Therefore, we will analyses two examples, fuel element vibrations in a flow with phase change and vibration induced by the leak flow in a BWR's jet pumps.

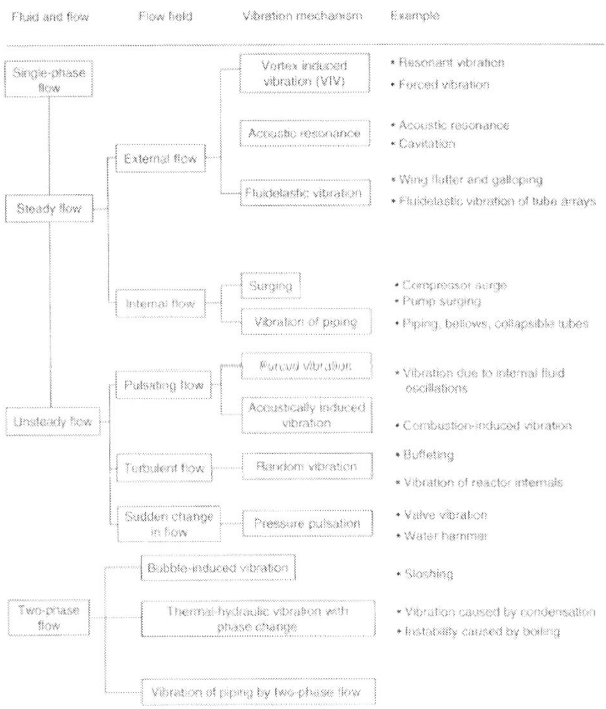

Figure 6. Vibration mechanism.

In the latter case, the jet pump FIV can be modeled as a vibration induced by the external axial flow. This is so because the leak flow through the Slip-Joint is deemed to be external to the mixer, as can be seen from Figure 7.

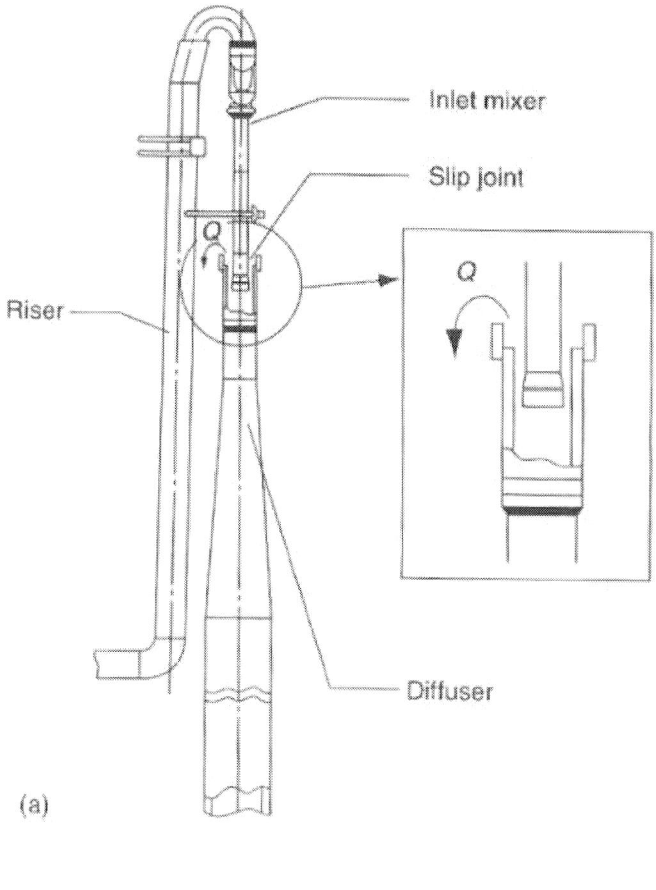

Figure 7. JetPump and Slip joint leak.

Thus, the model will be an axial flow cylinder, as described by Chen and Wambsganss (1970). The turbulent flow in the exterior, axial fluid gives rise to uneven pressure distribution on the outer wall of the pipe. Thus, on lacking balance it possesses resulting transverse forces. Moreover, as it is neither a permanent nor a stationary situation, the continuous change in pressure distribution over time, in space and in axial length, produces vibrations. The situation described is that which is shown in Figure 8.

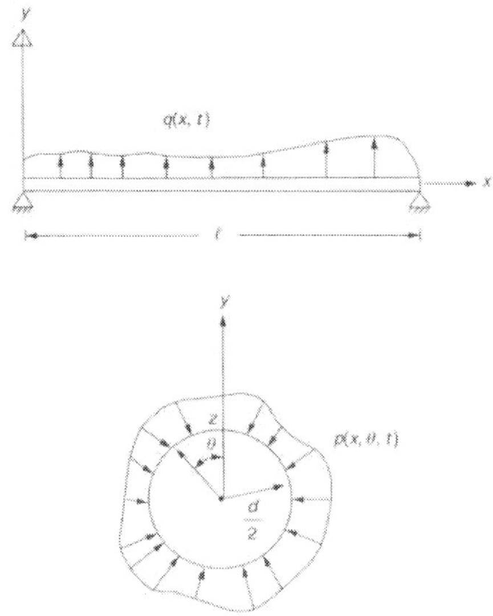

Figure 8. Pressure distribution in an external flow pattern.

TheCorcos (1963) model is used to analyses this type of situation. The model is expressed by:

$$\psi_{pp}(w, z_1, z_2, \theta_1, \theta_2) = \phi_{pp}(w) \cdot A\left(\frac{w|z_2 - z_1|}{V_c}\right) \cdot B\left(\frac{w \cdot D \cdot |\theta_2 - \theta_1|}{2 \cdot V_c}\right) \cdot e^{iw\frac{w|z_2 - z_1|}{V_c}}$$

where ψ pp is the cross-spectral density of the pressure field, φ pp is the pressure power spectrum at the point, A and B are spatial functions and V c is the convection speed.

Φ pp value is given by:

$$\phi_{pp}(fr) = \left\{ \begin{matrix} 0.272 \cdot 10^{-5} / fr^{0.25} \quad si fr5 \\ 22.75 \cdot 10^{-5} / fr^{3} \quad si fr5 \end{matrix} \right\}$$

where f r is the reduced frequency, which is given by the following, where Dh is the hydraulic diameter:

$$fr = \frac{f \cdot Dh}{V} = \frac{w \cdot Dh}{2 \cdot \pi \cdot V}$$

The convection speed of the limit layer is given by:

$$\frac{V_c}{V} = 0.6 + 0.4e^{\left(-2.2 \cdot \frac{w \cdot \delta}{V}\right)}$$

where

$$\delta = \frac{Dh}{2(n+1)}$$

$$n = 0.125 \cdot m^{3} - 0.181 \cdot m^{2} + 0.625 \cdot m + 5.851$$

$$m = \log\log(\mathrm{Re}) - 3$$

Lastly, the vibration amplitude is given by the approximations:

$$y_{rms} = \begin{cases} V^{1.5} si fr 0.2 \\ V^{2} si 0.2 fr 3.5 \\ V^{3} si 3.5 fr \end{cases}$$

These equations result in the generation of transversal forces, which move the mixer parts, the elbows and the riser. Moreover, as a result of its geometrical configuration, mode 3 vibration causes damage as a result of the stress at the first fixed point of the system, which is the joint between the riser and its support. Indeed, it is this stress that provokes breakages. A detail of the turbulence associated with the simulation of slipjoint leak can be seen inFigure 9.

In general, the Païdousis (1973) model is the equation used for industrial settings.

As far as FIV caused by external flow with phase change is concerned, its study and development can basically be put down to the nuclear industry, after the development of boiling water reactors. At such plants, boiling occurs on the external part of fuel rods in the upflow. This study was later extended to the steam generator pipes of pressurised water reactors.Figure 10.

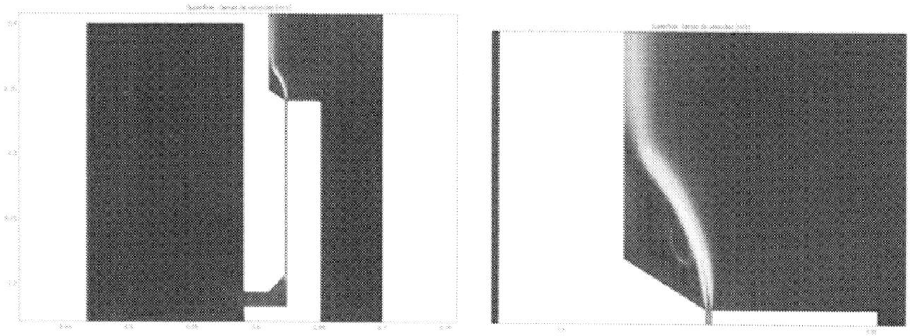

Figure 9. Simulated Slip joint leak and vortex induced.

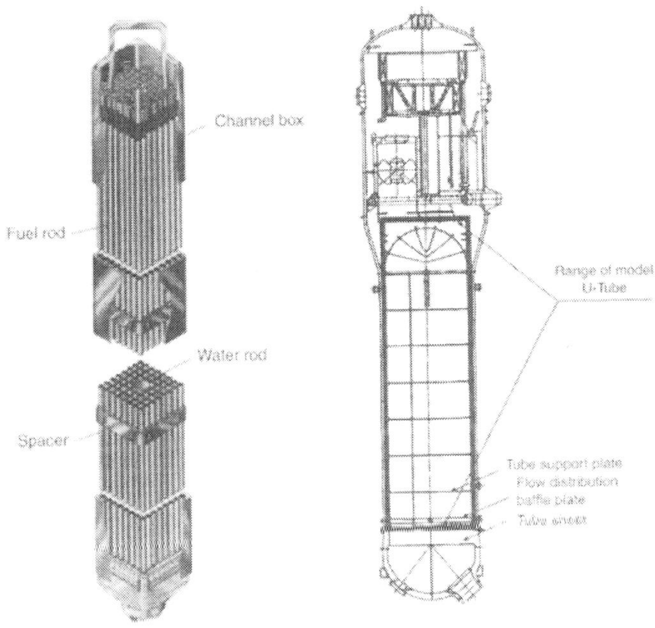

Figure 10. Fuel Bundle and Steam Generator.

Both studies reached the same conclusion, namely that the vibration amplitude is proportional to the mass up flow, but decreases with pressure. As far as fluid quality, or the void ratio, is concerned, two peaks can be seen, 0.1-0.25 and 0.4-0.5, which suggests that there is great dependence between vibration amplitude and the void ratio, or fluid quality.

On a BWR (Figure x) type fuel bar, it was established that the pressure power spectrum of the excitation force is proportional to V1.56-2.7, while the amplitude is proportional to V0.78-1.35. These values were determined in a rod test carried out by Pettigrew and Taylor [13] under the following conditions:

Pressure: 2.8-9 MPa

Mass flow: 0-4600 Kg/m2s

Power: 1-1000 Kw

Quality: 0-0.25.

Saito (2002) came to the same conclusions in another test under different conditions.

FLUID INDUCED NOISE (FIN)

In this section, we will take a look at the noise generated by a turbulent flow. Throughout this chapter and the following sections, fluid dynamics is based on movement equations, with speed as the unknown, essential analysis value.

For this particular case of noise induced by turbulence, a change of mindframe is called for in the sense of assessing the fluid in terms of pressure, and even density. Original FIN theory and expressions are given by the Ligthhill aeroacoustic model (1952/1954), as indicated below:

∇p2−1c20∂2p∂t2=−∂2Tij∂xi∂xj

Tij=ρvivj

Where *T* ij is the turbulent stress tensor. Whenever the fluid can be compressed, pressure variation is accompanied by density variation, the expression of which is:

*∂2ρ∂2t−c*0*∂2ρ∂xi∂xj=ρ*0∂2vivj*∂xi∂xj*

Thus, turbulence collaterally generates pressure and density variation in the fluid. By means of the turbulent stress tensor the turbulence produces variations in pressure and density. The former cause the noise, and as such, are deemed to be sound sources.

There are many practical applications of the analysis of turbulence generated noise. Two particularly curious, albeit useful ones, have solved some serious problems. The two situations in question are:

- Determination of leaks through the seat of safety relief valves from the outside by means of non-intrusive techniques.
- Element breakage due to resonance frequencies.

The first of the above situations has been used to detect safety relief valve leaks in BWR nuclear plants. The theoretical principle employed is that the seat leak flow produces turbulence which in turn generates a characteristic sound. The conceptual model is similar to the one used by Van Herpe and Creghton (1994), in which they model the fluid flow through a conduit with a restriction inside. These researchers came to the conclusion that the acoustic power, dB, is proportional to the speed in the conduit and to the passage diameter determined by the restricting element. Consequently, an SRV leak can be detected by means of sound and its register, but also by its evolution. The latter claim is based on the tests done by EPRI on this particular study NP-2444-SY. In this study, the frequency band is established at which the leak is best detected, 40-55 KHz, as well as the complete detection interval, 30-60KHz. Likewise, the author holds that regular monitoring can establish drift and trend patterns, the development mathematical law of which is that determined by Van Herpe y Creghton (1994).

The other case refers to the catastrophic breakages of steam dryers at BWR plants due to an acoustically sourced resonance. (Figure 11).

After the failure of the Quad cities dryer, a study was carried out on the loads to which it was subject. In none of the cases did the operating loads justify the breakage, or degradation, of the component. Consequently, a study was made of the vibratory or pulsatory phenomena, by measuring the vibrations at the component and the passage of pressure along the main steam lines. Acoustic excitation coming from the main steam lines is the current model. (Figure 12).

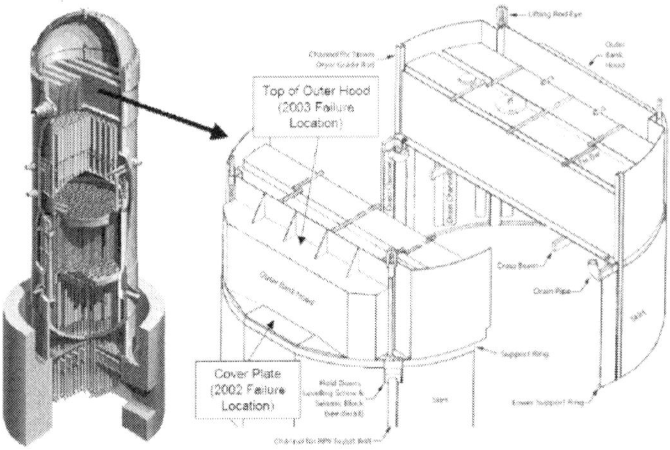

Figure 11.BWR vessel and Steam dryer detail. (Cracked areas)

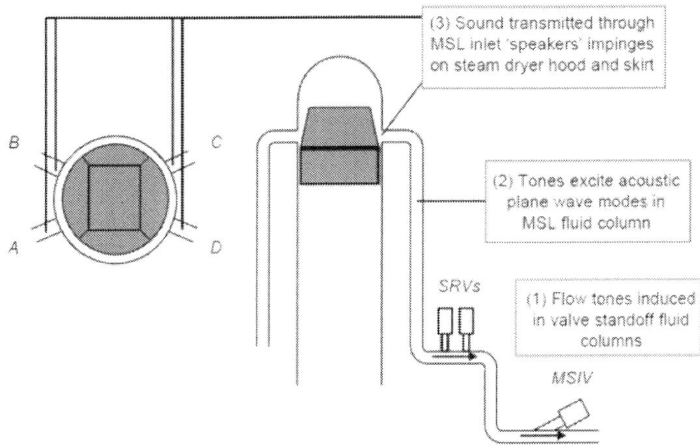

Figure 12.MSL, SRV's and the main characteristic of the sound transmission.

Acoustic excitation is produced in the main steam lines as a result of the turbulence generated at the SRV connections, instrumentation connections, equalising header, etc... (Figure 13)

This turbulence, which in general, manifests itself as vortices or eddies will, in the Von Karman sense, give off a determined frequency. Moreover, the lower scale turbulence will have its own frequency in accordance with the Lighthill model. Thus, the pressure waves that are generated with a defined frequency will constitute an external dryer load that increases its total dynamic load. (Figure 14).

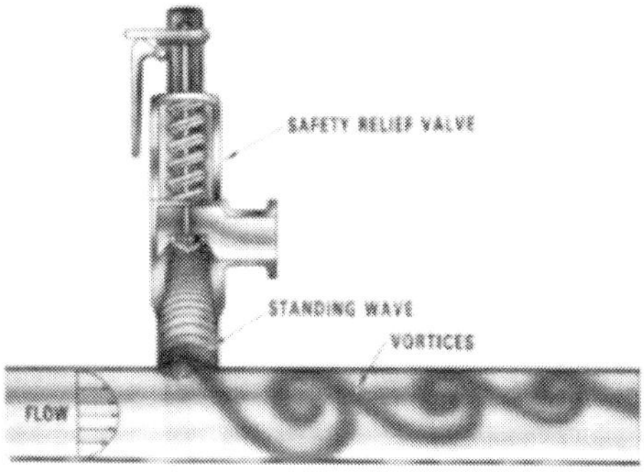

Figure 13. Singing branch. Vortex Sheading.

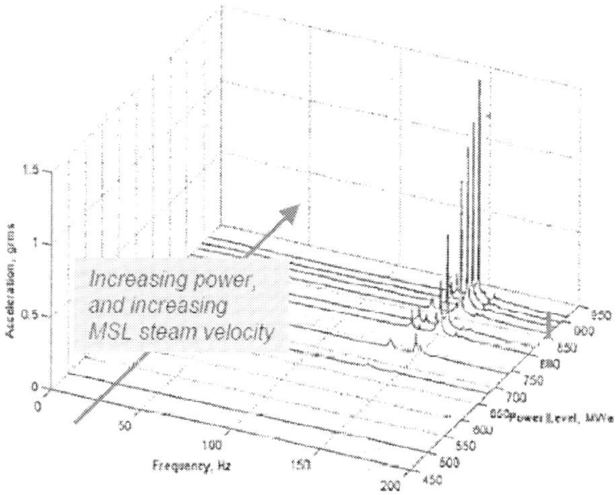

Figure 14. Frequency dependency of the power and acceleration of the induced load.

Moreover, this load with be pulsatory in nature and will have a determined frequency, thus, if it coincides with a natural dryer frequency, the pressure effects are amplified, exceeding the resistance of the material, in some cases breaking, in others causing cracks.

As can be seen, there are frequencies in Figure that coincide with certain natural ones of parts of the dryer (Figure 15). Consequently, there will certainly be damage. The only variable will be time and the power of the vibration due to the reactor power. Notwithstanding, the solution to the problem is to be found at the same source as the excitation, namely, by studying the turbulence. It has been proven that by changing the geometry of the SRV connection, this changes the spectrum frequency, and consequently, the effect on the dryer. (Figure 16).

In line with the practical focus followed in this chapter, there are reduced analytical studies that make it possible, in a T-type geometry, to predict and check the existence of acoustic excitation applied to steam pipes.

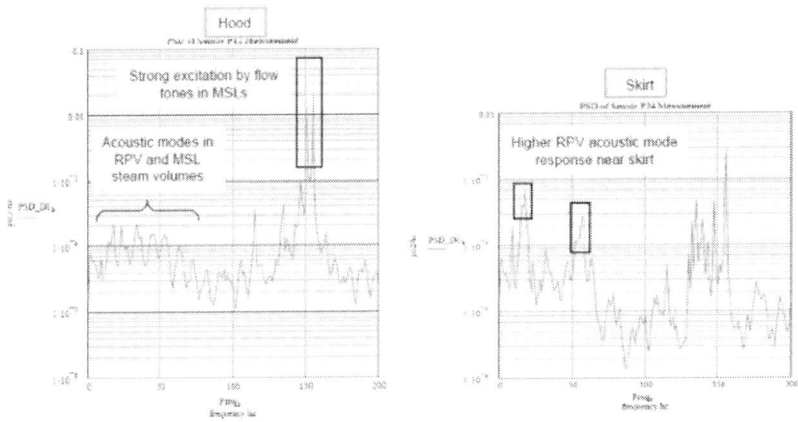

Figure 15. Natural frequencies of the steam dryer Hood and skirt.

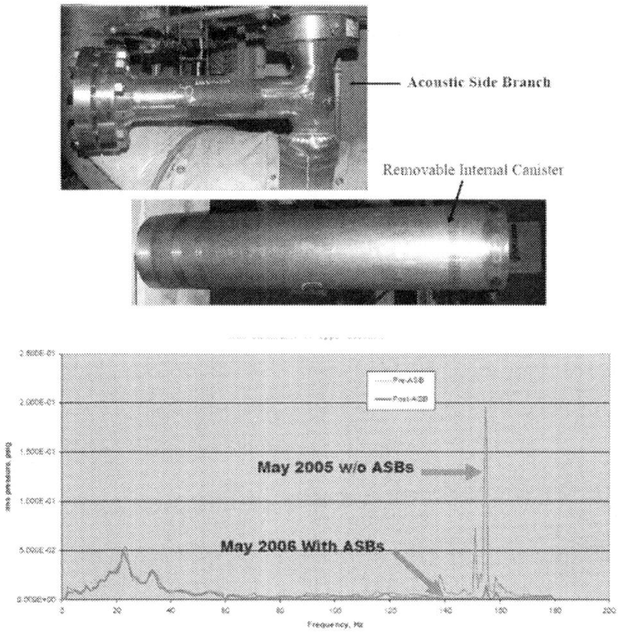

Figure 16. Acoustic side branch, and the effect on the pressure, or load.

The mathematical expression is:

$$S0 = a \cdot N \cdot (d+r)4 \cdot L \cdot V0$$

where:

$S\ 0$ is the Stroudhal number;

α is the speed of sound in a fluid medium;

N takes on value of 1 and 3;

d is the branch diameter;

r is the branch curve radius at its connection to the main line;

L is the branch length;

$V\ 0$ is the main line speed.

The assessment is carried out in accordance with the Strouhal number, defined by the equation x, if the value lies within the 0.25-0.60 interval, for which it can be said that there is a big likelihood of acoustic excitation appearing. The frequency estimate can be approximated by the following function:

$$S0=fs\cdot d/V0$$

Fluid induced instabilities.

In this section, we are going to deal with fluid instabilities, though more specifically with that which makes the flow bisTable, given that this represents a transition within the turbulent system. Notwithstanding, what follows is a brief description of other instabilities about which references and studies abound. Turbulent flow, as has been above, is a generalised process, whereas instability is different. This is an unexpected situation, and one which, in principle, should not be happening. Quite often it is associated with the local formation of turbulence in a laminar, ordered flow. The most widely known instabilities model and study turbulence of this type of situation, but there are other instabilities in a turbulent flow that cause changes to the flow pattern. As with turbulent flow and its transition, instabilities possess factors that give rise to same, which are:

- Fluid-structure interaction: Remodels the flow
- Von Karman
- Stroudhal
- Density or viscosity variations
- Raleigh-Taylor
- Plateau-Railegh
- Ritchmyer-Meshkov
- Speed difference
- Kevin-Heltmotz.
- Increase in turbulence intensity
- BisTable flow.

All of these instabilities exert different effects on the structure that is either immersed in them, or that contains the flow, such as vibrations, noise or change in process flow values. The latter, is of itself only an instability, albeit a very perculiar one, thus it will be dealt with separately. The others, given their spatial characteristics, have been dealt with separately in Sections 4 and 5.

Presented here is an important advance on the investigation of the bisTable flow phenomenon inside a boiling water reactor

(BWR) vessel. The study of the flow time series concludes that the phenomenon of the bisTable flow is a particular case of transitions induced by turbulence. Afterwards the phenomenon is identified, simulated and the model is validated. The real behaviour is reproduced using the mathematical model, which has given excellent results. Therefore the bisTable flow is an instability that comes by transitions induced by turbulent flows and vortex.

During this analysis a new technique was used to characterize the phenomenon. The new technique was the Hilbert´s transform. The conclusion that came from this last study is that:

- The bisTable flow is non linear and non stationary
- The bisTable flow is produced when the plant is near the coast down and the flow is the maximum
- There is a relationship between the average value of the flow and the bistability.
- The bisTable flow has two attractors structure.

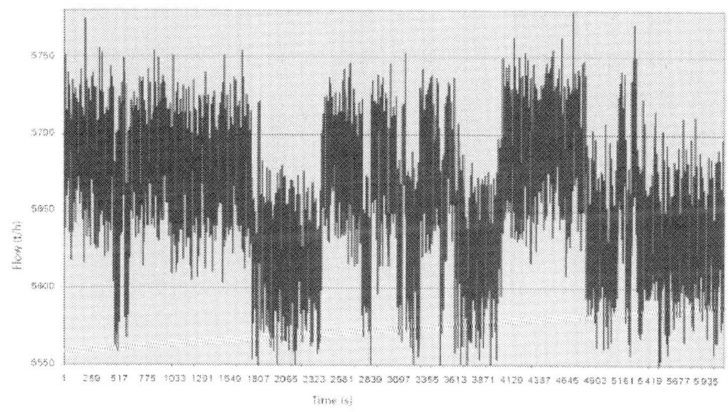

Figure 17. Flow rate time series of in the recirculation loop. BisTable event.

One of the most absurd process features is that the white noise can induce order in a system that is non linear and non stationary, is not in equilibrium. That is, a chaotic system can be ordered by itself. This is called vibration resonance (Horsthemke, 2005), with transitions induced by noise. In this case the word noise and turbulence will be similar. So this phenomenon is modeled by the Laugevin´s equation (56).

$$x^{\cdot}i=f(xi)+g(xi)\cdot\varepsilon(t)+D/2\cdot d\ \Sigma(xj-xi)+A\cos(w\cdot t)+B\cos(\Omega\cdot t)$$

Where:

x(t)i is the state of the i-esim oscillation.

ε(t) it is a gaussian noise zero-averaged an non correlated.

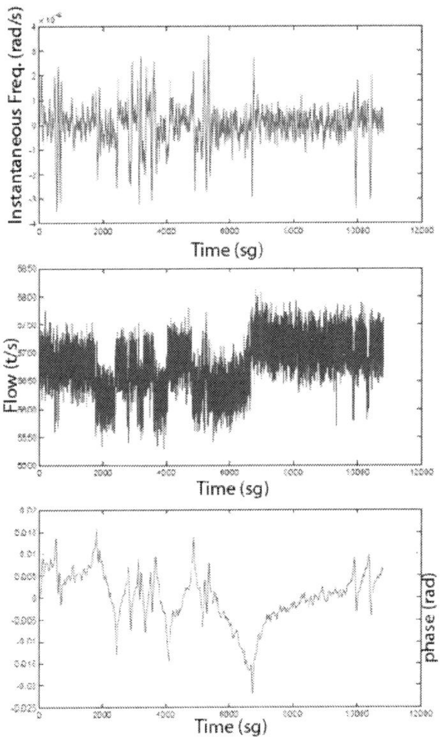

Figure 18. Results of the Hilbert´s Transform. Real Time series (central), Phase (bottom), Inst. Frequency (upper)

The last terms in the equation, represent the external and periodic forces represented as low frequency and amplitude A, and another of high frequency with amplitude B.

For the case under study the expressions of the functions g(x) and f(x) are:

G (x) = K1x

F (x) = − K2x−K3x3(7)

For the above-mentioned the equation of the flow would be in the way:

$$\dot{x}_i = \varepsilon \cdot K_1 \cdot x - K_2 \cdot x - K_3 x^3 + \sqrt{\eta} \cdot \zeta$$

Resuming:

$$\dot{x}_i = x(\varepsilon \cdot K_1 - K_2) - K_3 x^3 + \sqrt{\eta} \cdot \zeta$$

Simplifying:

$$\dot{x}_i = x \cdot K_1' - K_3 x^3 + \sqrt{\eta} \cdot \zeta$$

Where:

x(t) is the flow and ζ is the zero averaged and non-correlated noise white. The parameters K1′ and. K3 will be calculated to determine the times in those that, statistically, the system is in each value, or also called escape times.

To determine the values of the K´s factors it must be consider that:

The averaged value of the flow is 5650 t/h and if there´s no noise or turbulence the results of the equation (61) are zero.

The averages deviations of the average value must be +50 and -50 t/h.

The values of the parameters expressed in the equation (61) and that are coherent with the previous premises are given by the Table 2.

Table 2 . Values of parameter to equation (58).

Parameter	Value
K1`	0.0052
K2	8e-11

η	0.1797
Desv. Typical Noise	40

The realization of the equation (61), gives results such the follows, showed in Figure 19 and 20.

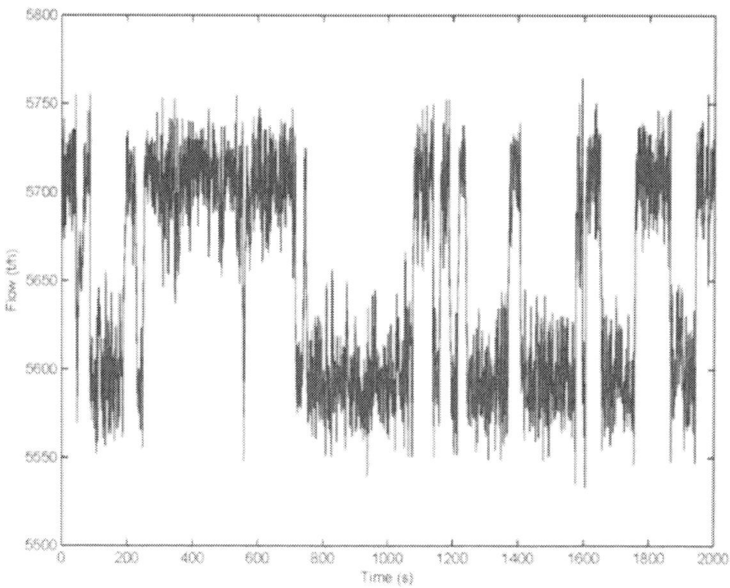

Figure 19. Realization of the equation (61) with the parameters of the Table 2.

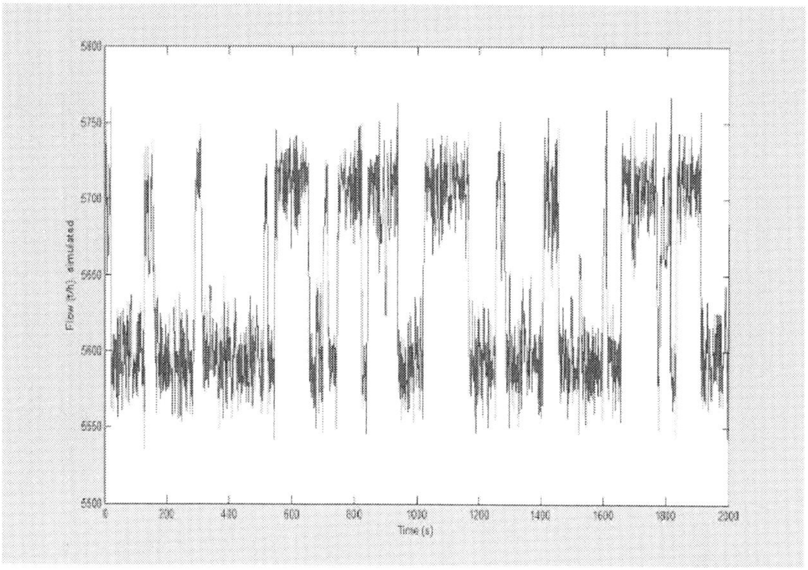

Figure 20. Realization (2nd) of the equation (61) with the parameters of the Table 2.

The previous graphics are two drawings of the flow value in sets of 2000 samples, in order to have a more representative graphic.

The difference between the simulation and the real time series is due to the measurement method that adds a new noise to the signal, but that noise is not important to the phenomenon. The flow is measured by the difference of pressure in an elbow, so in this method the centrifugal force has to be measured and then the mass flow.

The validation is based in two steps. The first step is the generation of the histogram, autocorrelation graphic and the spectral density graphic (Figures 21, 22 and 23) to compare with the results showed in the characterization previous work done by Gavilán (2007). And the second is the evaluation of the phase diagram given by the Hilbert´s transform over the simulated data and the comparison with the real data phase

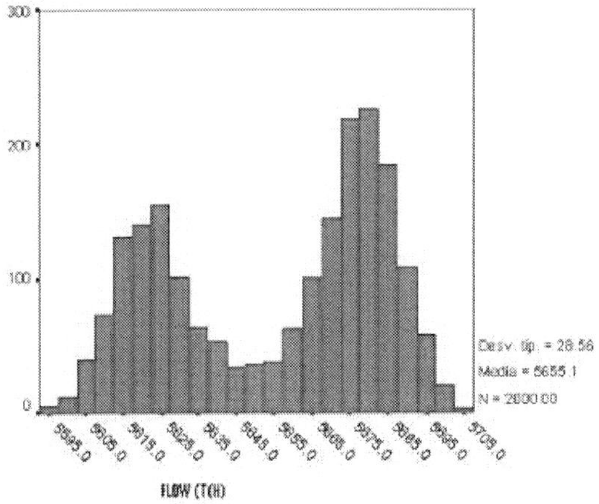

Figure 21. Simulated time serie Histogram. (2000 seconds)

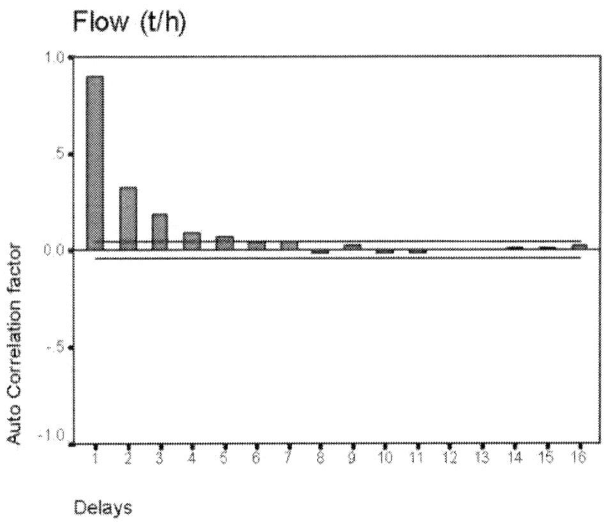

Figure 22. Simulated time serie Autocorrelation factor

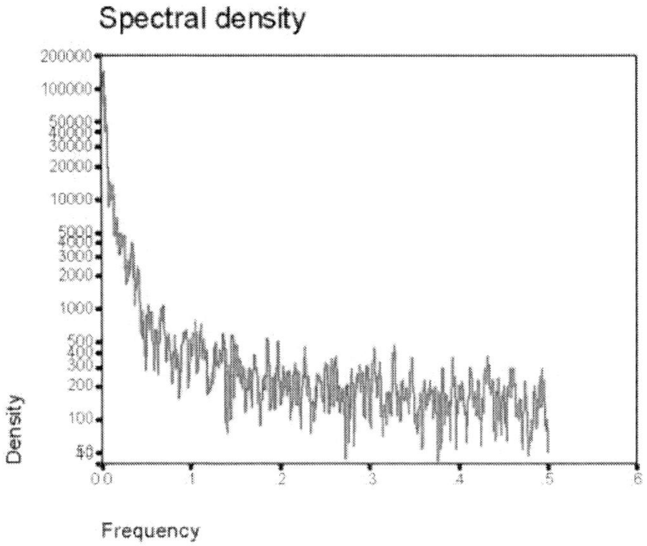

Figure 23. Simulated time series Spectrogram.

The statistic parameters of the simulated series, (from Figures 21,22 and 23) are coherent with the results found by Gavial (2007) in the characterization of the bisTable flow signal.

To conclude the validation method, is necessary to check if the simulated time series have the same characteristics as the real one in the Hilbert´s sense. So to compare is necessary to repeat the Hilbert´s transform but now to the simulated time series. The results are showed in theFigure 24.

In the Figure 24, y can be observed the structure of two attractors in the phase, and in the instantaneous frequency. The law for the transitions between onc and the other attractor, in the phase space, is an arcos (K (t)) and the instantaneous frequency is the derivate of it. If we compare theFigures17, 18, 19, 20 and 24 can be concluded that the simulated series adjust fairly to the real one. The adjustment appears very qualitative but must remember that the relationship in statistical.

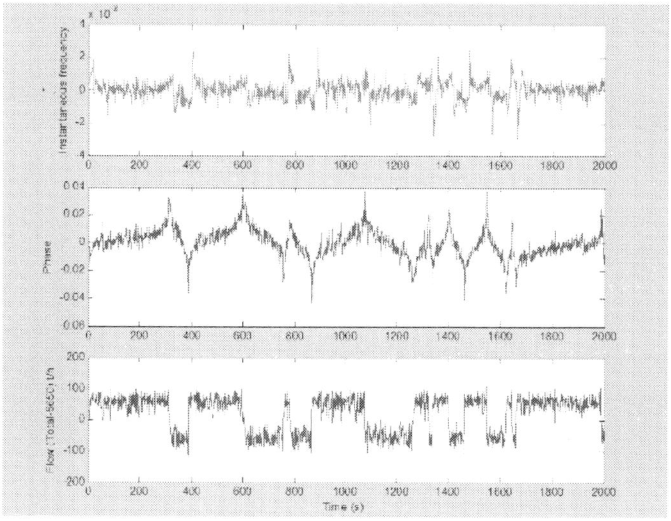

Figure 24. Results of Hilbert´s transform to simulated data.

The characterization of the bisTable flow results in, that is nonlinear and non-stationary process. The bisTable flow can be evaluated and characterized, properly using the Hilbert´s Transform.

The bisTable flow appears under some operative conditions without change in the values of control variables, so is associated to an internal variation. The variation is due to the turbulence that is a self-oscillatory process in a continuous medium. The development of a turbulence event is the birth of an attractor in the space of the phase. The turbulence is a convective process because it doesn´t increase indefinitely but rather it evolve to water below. All those conclusions about the turbulence justify that the flow change the average value of the speed and then the value of the flow taking one of the two values of the attractor.

The turbulence is an unstructured process, because there´s no pattern in it. The Fourier transform gives poor results when is used so that´s the reason why the turbulence is associated to a noise, and sometimes to a white noise. The phenomenon of bisTable flow is a turbulence induced transition process, because the reason of the jumping between two flow values is the turbulence. Because turbulence and noise are physically the same, mathematically the phenomenon can be simulated with a noise-induced transition model.

The simulation gives us, a low error simulated time series that had the same statistical parameters than the original. And the other idea is that the bus Table flow only appears in a finite region of the flow space, because there´s only a set of values of the turbulence and flow that induce the bisTable behavior. In the same idea, the noise (turbulence) has only one value to induce the bisTable flow. The characteristics parameter of the noise is the average that must be zero, and the standard deviation that has a relationship with the value of flow, and only in a small interval of values produce the bus Table flow (Table 3,Figure 25). The relationship between the noise standard deviation and the flow depends in every installation because depends on the hardware but in this case can be modeled by the equation (62).

$$StdDev(F) = 30.02 + \frac{A}{\frac{B^2}{Flow(t/h)^2} - 1}$$

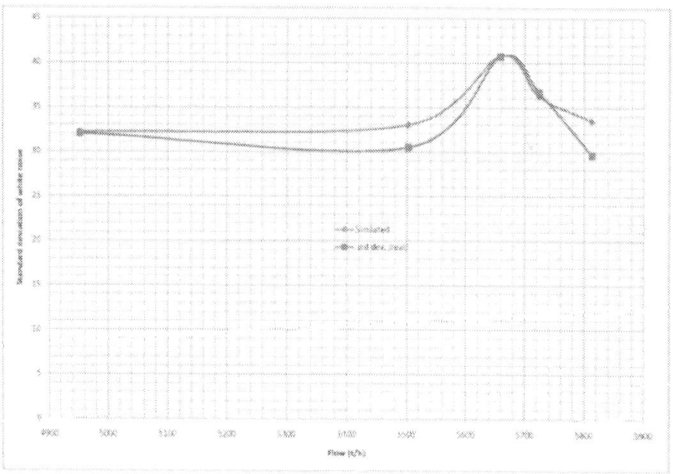

Figure 25. Relationship between average flow and noise typical deviation.

Table 3.Parameters of several intervals of flow and conditions.

Interval Time	Min. Flow (t/h)	Max. Flow(t/h)	Average Error	Typical Deviation	R2 adjusts lineal	Bistability
300-1200	4423.9	5479.2	4.17	32.01	0.9844	No
1500-3000	5323.3	5682.9	-41.01	30.47	0.7952	No
3001-10000	5492.7	5825.7	0.091	40.71	0.5011	Yes
10001-13000	5579.9	5867.7	0.055	36.72	0.2972	Yes, smaller
13001-15000	5695.3	5929.5	0.0455	29.63	0.1482	Residual

CONCLUSION

This chapter has shown that turbulence exercises several effects in the industrial setting, and except for those of diffusion, or mixture, all the others are both harmful and give rise to problems. Given its variety and applicability, it is not easy to discover the reason for a component failure. Sometimes, only an in-depth analysis of the problem from the perspective of turbulence can serve to clarify a case that defies normal analyses.

Therefore, a thorough grounding in fluid mechanics is needed to solve the problems related to turbulence. Furthermore, knowledge of computer fluid flow modeling tools, which involves so-called Computational Fluid Dynamics (CFD), represents a great advance in the knowledge of the turbulent behavior of fluids in movement. Notwithstanding, all is not science and mathematics, given that turbulence is chaos, and therefore, there will always be a certain druidic facet in those professionals given to working in this field.

As an end to this chapter, we provide a summary Table (Table 4) that can serve as a guide to locating what effect may be producing the turbulence, with the aim in mind of better focusing the subsequent analyses.

Table 4. Summary of fluid type and effects.

Fluid type // condition	Turbulence effect	Affected area
Single-phase // Saturated fluid	Cavitation	Venturi tubes and geometry changes
Single-phase fluid	Instability	Connection geometry
	Noise // Acoustic Excitation	Connection geometry
Fluid structure	Vibration	Elements submerged in the current
	Noise // Acoustic Excitation	Elements submerged in the current

Lastly, a look at the use of CFD´s. The use of this type of tool does not ensure the success of the study, given that employing it a preliminary analysis needs to be done and the following defined:

- Model type (LES, RANS, DNS…)
- Minimum modeling scale (Kolmogorov)
- Analysis type (permanent or provisional system)
- Variable to be simulated (density, pressure or speed)

Otherwise, the results from the computer will be nothing more than a nicely coloured drawing, without any coherent information.

"Linearity is an idea sought after, yearned for and forced by the human mind. It is a reflection of our condition as animal. Only by accepting, assimilating and understanding the chaos of the world that surrounds us, will we truly ascend to the rational condition."

Notes

[1] -These are referred to as coherent because the vorticity is concentrated and the fluid flows around as if it were a solid obstacle. It keeps its shape for longer than a single rotation.

REFERENCE

1. United Stated Nuclear Regulatory Commision. 1986 Anomaluos behavior of recirculation loop flow in jet pump BWR plants, Information. notice 86-110 86 110 . Washington.
2. A. Nuñez, E. Carrera, G. Martinez-Mendez, Paredes. Espinosa, 2006

Análisis del fenómeno de flujo biesTable de la central Nucleoeléctrica de Laguna Verde. Proceedings of the 2006 Congreso Internacional Buenos Aires. LAS/ ANS.

3. C. J. Gavilán Moreno, 2007 El flujo biesTable en lazos de recirculación de una central BWR. Analisis y caracterización. Proceeding of the XXXIII Reunión anual de la Sociedad Nuclear Española.
4. C. J. Gavilan Moreno, 2008 Analysis and simulation of the flow signal in the recirculation loop of a nuclear power station during a bisTable flow event. Nuclear engineering and Design. 238 10 2754 2760 .
5. C. J. Gavilán Moreno, 2009 Hydraulic study on recirculation loops using computational fluid dynamics (CFD). Design optimization and turbulence reduction. Nuclear Engineering and Design, 239 3 434 441
6. N. E. Huang, 1996 Introduction to the Hilbert-Huang transform and its related mathematical problems. Goodard Institute for Analysis dates. NASA.
7. S. Ziada, S. Shine, 1999 Strouhal numbers of flow excited acoustic resonance of closed side brances. Journal of fluids and structures. 13 127-142.
8. R. M. Baldwin, H. R. Simmons, 1986 Flow induced vibration in safety relief valves. Journal of pressure vessel technology. 108 267 272 .
9. O. C. Zienkiewicz, R. L. Taylor, 2004 El método de los elementos finitos. 3 dinámica de fluidos. Ed. CIMNE.
10. K. Morgan, E. Oñate, J. Periaux, J. Peraire, O. , 1993 Finite element in fluids. New trends and applications. Ed. CIMNE.
11. M. Lesieur, 2008 Turbulence in fluids Ed. Springer.
12. Horsthemke & Lefever 2005 Noise Induced Transitions. Ed. Springer series in Sinergetics.
13. Frank. M. White, 2008 Mecánica de fluidos. Ed. McGraw-Hill.
14. L. D'Agostino-Salvatore, Eds 2007 Fluid Dynamics of Cavitation and Cavitating Turbopumps. Ed. SpringerWien NewYork.
15. Per. Persson-Olof, 2002 Implementation of finite element based Navier Stokes Solver. 2094 Project.
16. F. Bigersson, S. . Finnveden, G. Robert, 2004 Modelling turbulence induced vibration in pipes with a spectral finite element method. Journal of sound and vibration. 278 749 772 .
17. A. N. Kolmogorov, 1941 The local structure of turbulence in incompressible viscous fluid for very larger Reynolds numbers. Proc. R. Soc Lond. A. 434 9 13 .
18. A. Roshko, 2000 On the problem of turbulence. Current Science, 76 6 834 839 .

19. J. Dassé, S. Mendez, F. Nicoud, Large-Eddy Simulation of Acoustic Response of a perforated plate. American Institute of Aeroanutics.
20. F. Van Herpe, D. G. Crighton, 1994 Noise generation by turbulent flow in ducts. Journal de Physique III, 4 947 950 .
21. P. Davies, K. R. Holland, 2004 Flow Noise generation in expansion chambers. Procc of the Institute of Acoustic. 26 Pt. 2, 206 213 .
22. H. Hamakawa, et al. 2006 Effects of flow induced Acoustic resonance on vortex shedding form Staggered tube banks. JSME International Journal. Series B, 49 1 142 151 .
23. S. S. Chen, M. W. Wambsganss Jr, 1970 Response of a flexible rod to near field flow noise. Proceedings of the conference on flow induced vibrations in reactor system components. ANL-7685 Argonne National Laboratory.
24. G. M. Corcos, 1963 Resolution of pressure in turbulence . J. Acoustic Soc. Am. 35 2 192 199 .
25. M. P. Païdoussis, 1973 Dynamics of cylindrical structures subjected to axial flow. Journal of Sound and vibration. 29 3 365 385 .
26. N. Saito, et al. 2002 BWR 9x9 Type Rod Assembly Thermal Hydraulic Test (2)- Hydraulic Vibration Test. 10th International Conf. on Nuclear Engineering. ICONE10 22557 .
27. M. J. Lighthill, 1952 On Sound Generated Aerodynamically. I. General Theory, Proc. R. Soc. Lond. A 211 564 587 .
28. M. J. Lighthill, 1954 On Sound Generated Aerodynamically. II. Turbulence as a Source of Sound , Proc. R. Soc. Lond. A 222 1 32 .

Chapter 6

WIND VELOCITY DECREASING EFFECTS OF WINDBREAK FENCE FOR SNOWFALL MEASUREMENT

Ki-Pyo You and Young-Moon Kim

Department of Architecture Engineering, Chonbuk National University, 567 Baekje-daero, deokjin-gu, Jeonju-si, Jeollabuk-do 561-756, Republic of Korea

ABSTRACT

Meteorological observatories use measuring boards on even ground in open areas to measure the amount of snowfall. In order to measure the amount of snowfall, areas unaffected by wind should be found. SSS We performed a computational fluid dynamics analysis and wind tunnel test, conducted field measurements of the type and height of the windbreak fence, and analyzed the wind flow inside the fence. The results showed that a double windbreak fence was better than a single windbreak fence for decreasing wind velocity. The double fence (width 4 m, height 60 cm, and fixed on the bottom) has the greatest wind velocity decrease rate at the central part of octagonal windbreak.

INTRODUCTION

Fresh snowfall is defined as new snow covering an even plane. The depth of snowfall is defined as the increment of the snow layer cover during the time of measurement. The snow depth is generally measured each day and the snowfall is reported in centimeters per day. The depth of fresh snowfall on open ground can be measured directly using scaled rulers or a snow ruler. The depth of snow covering the ground, or accumulated on the ground, is measured by inserting a scaled rod vertically into the ground. However, snow covering the ground may come into the measurement plate by wind. Another limitation is that the snow ruler might meet an ice layer, rather than the ground. Thus, it is difficult to obtain a representative height measurement in an open area. We need to make sure that the total depth, including the ice layer, is measured and we need to average multiple measurements from each observation. The amount of snowfall measured in open areas is greatly affected by wind. Piles of snow formed by frequently blowing wind are scattered, resulting in some difficulty in evaluating the exact amount of snowfall. Current studies have focused on diverse equipment to measure the exact amount of snowfall. In addition, snowfall equipment is installed in areas where snowfall is not particularly influenced by wind. However, there is no research on wind control at the point where the amount of snowfall is measured. There has been much research undertaken on windbreak fences. In 1971, Plate [1] subdivided diverse air current areas in relation to turbulence boundary layers near windbreak fences. Based on the results of research into wind decrease areas at the back sides of fences, a diversity of windbreak fences has been used. Perera [2] and Ranga Raju et al. [3] reported that the wake phenomenon disappeared on the back side of the fence when a windbreak fence with a porosity of 30% was installed. Judd et al. [4] tested wind tunnels on the flow around single and multiple porous windbreaks sheltering a model plant canopy. They found that a windbreak has effects on the flow within the sheltered canopy, both upwind and downwind of the break. There is acceleration within the canopy just upwind of the break, in contrast with a deceleration above the canopy. Lee and Kim [5] and Park and Lee [6] performed a wind pressure experiment in a triangular open air storage yard, changing the height of the windbreak fence, porosity, and interval, and reported that their best results were achieved with a porosity of 30–40%. Research on windbreak fences has focused on changes in porosity rates that are aimed at decreasing the wind velocity and turbulence caused by back streams within the fences. However, research on windbreak fences and their use in measuring snowfall aims to decrease wind velocity and minimize the effects by wind when snow is piled. In order to achieve this goal, an evaluation was made using three methods. A decrease in wind velocity is considerably affected by a windbreak fence's height and width. In light of this, a computational fluid dynamics (CFD) analysis was conducted in order to determine its optimal form. A wind tunnel test was conducted in order to examine actual wind velocity and its turbulence-decreasing effect at the windbreak fence, using the CFD analysis results. Field measurements for analysis and verification of experimental results were made.

DATA AND METHODOLOGY

In order to determine the height of a windbreak fence and experimental wind velocity, the amount of snowfall and wind velocity was analyzed using data on weather conditions over five recent years (2008–2012). Data from weather stations located in western and eastern coastal areas (where heavy snowfalls occur) was the primary source. In the western coastal area, data from weather stations in Gochang (35°20'N, 126°35'E), Gunsan (36°00'N, 126°76'E), and Mokpo (34°81'N, 126°38'E) were used, while in the eastern coastal area, weather stations in northern Gangneung (37°80'N, 128°85'E) and Daegwallyeong (37°67'N, 128°71'E) were analyzed. Figure 1 shows the locations of the weather stations. Figure 2 shows the frequency of snowfall and the frequency of wind velocity over five recent years. In the western coastal area of the peninsula, snow falls more frequently than it does in the east, due to high atmospheric pressure from Siberia and the water temperature of the western region. Nevertheless, a large amount of snow falls in the east coastal area due to high atmospheric pressure from Siberia and the water temperature and mountain ranges. The average amount of snowfall in three areas, with a high frequency of snowfall in the west coast, was 3.6 cm, while the maximum amount of snowfall was measured at 24.5 cm at Gochang weather station, which is located inland. In 90% of the surveyed areas, the frequency of snowfall was less than

10 cm. In more than 50% of the surveyed areas, the frequency of snowfall was less than 2 cm. On the east coast, the average amount of snowfall in two areas where snow fell frequently was 8.75 cm and the largest amount of snow fell in northern Gangneung, which is located inland, at 77.7 cm.

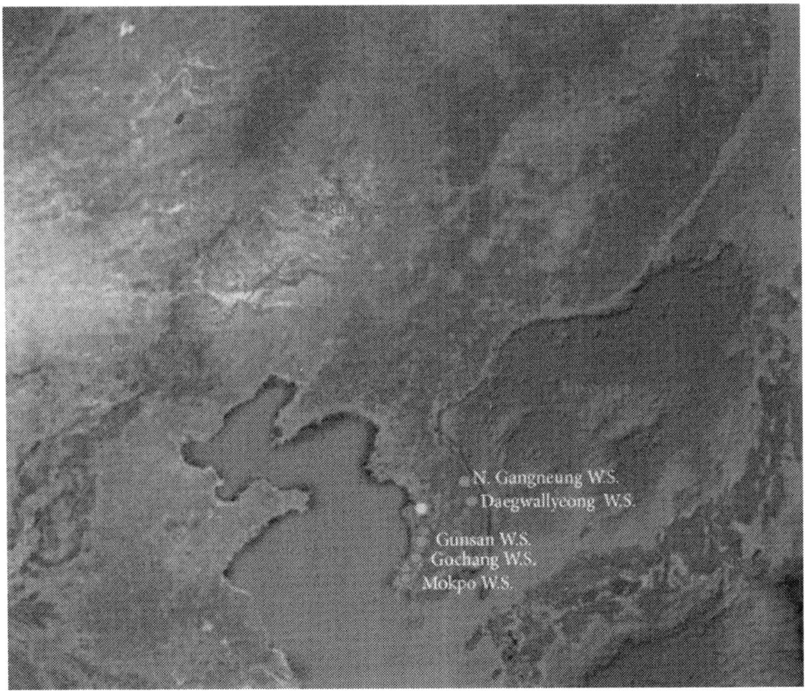

Figure 1: Location of weather station.

The areas where snowfall frequency was less than 10 cm accounted for 75% of the total frequency. The areas where the frequency of snowfall was smaller than 5 cm made up more than 57% of the total frequency. Regarding wind velocity distribution during snowfall, the average wind velocity of three areas in the west coast was 3 m/s and the maximum wind velocity was 11.2 m/s, at the Mokpo weather station. During snowfall, the frequency of wind velocity at less than 6 m/s amounted to 82% of the total frequency. The average wind velocity of two areas on the east coast was 2.5 m/s, with the maximum wind velocity, measured at Daegwallyeong weather station, being 12.4 m/s. During snowfall, the frequency of wind velocity at less than 6 m/s amounted to 96% of the total frequency. Using data from weather stations, the height of windbreaks for snowfall measurement was determined to range from 40 cm to 80 cm, and the experimental wind velocity was determined to be between 3 m/s and 6 m/s.

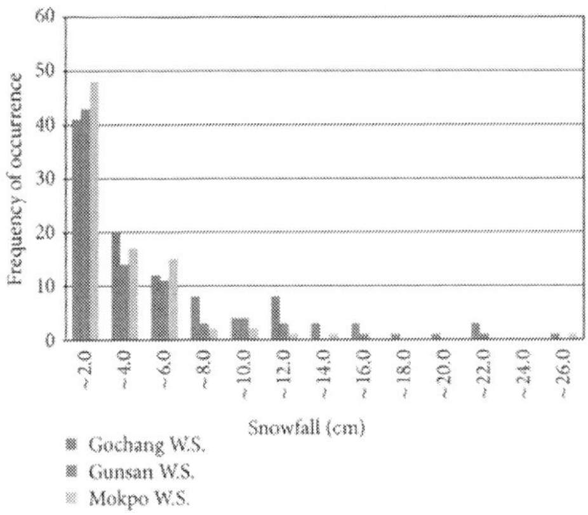

(a) Snowfall frequency of occurrence of the western coastal area

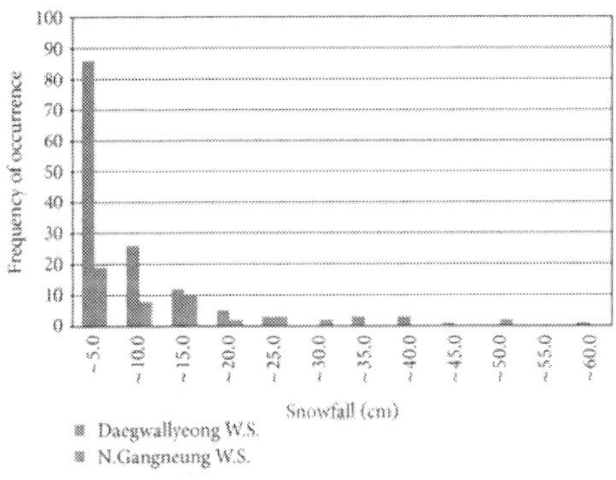

(b) Snowfall frequency of occurrence of the eastern coastal area

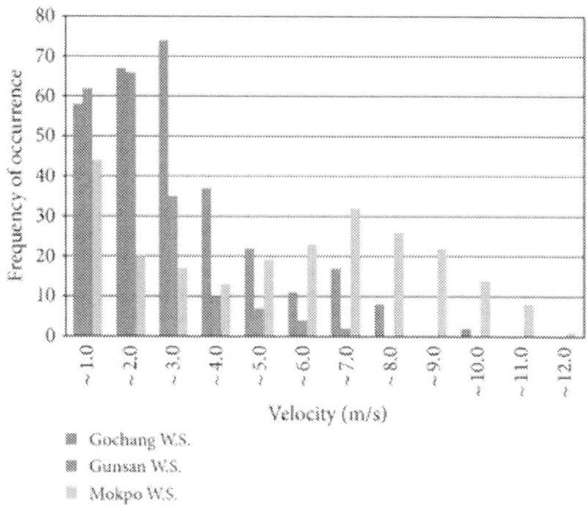

(c) Wind velocity frequency of occurrence of the western coastal area

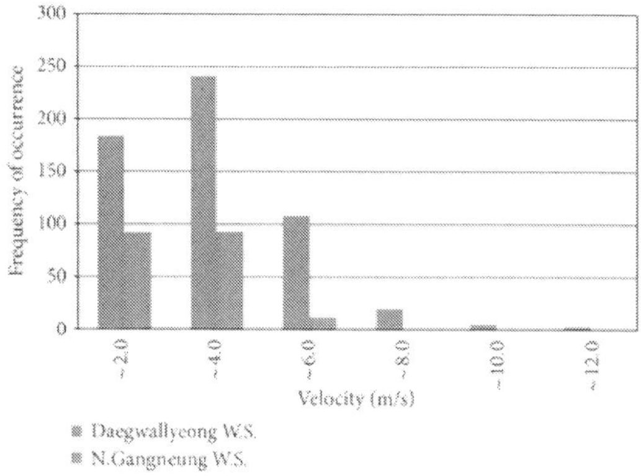

(d) Wind velocity frequency of occurrence of the eastern coastal area

Figure 2: Snowfall frequency of occurrence and wind velocity frequency of occurrence.

COMPUTER FLUID DYNAMICS

Geometric Model and Mathematical Model

The geometric model of the windbreak fences exhibits suburban conditions. The dimensions of the linear windbreaks and the windbreak fences according to the analysis model were $5\ m(x) \times 20\ m(y) \times 4\ m(z)$

and $10\,\mathrm{m}(x) \times 14\,\mathrm{m}(y) \times 4\,\mathrm{m}(z)$kε respectively, with the parallelepiped situated at the center. Figure 3 shows the analyzed domains. As the meshes for the analyzed domains, 0.5 mm triangular meshes were used. The number of meshes of the analyzed domains was between 1,100,000 and 1,500,000. Commercial software scStream V9.0 was used for the CFD simulation [7]. As a representative model that can simulate flow around the fences, the standard kε model was used. The standard kε model has excellent convergence properties and is mathematically simple. Its calculation time is also short. Several studies have used CFDs to model wind flow around buildings by solving mass and momentum conservation equations:

$$\begin{gathered} \frac{\partial \rho}{\partial t} + \nabla \cdot \left(\rho \vec{v}\right) = 0, \\ \frac{\partial \rho}{\partial t}\left(\rho \vec{v}\right) + \nabla \cdot \left(\rho \vec{v} \vec{v}\right) = -\nabla p + \nabla \cdot \left(\bar{\bar{\tau}}\right) \rho \vec{g}. \end{gathered} \tag{1}$$

Here, ρ is the density

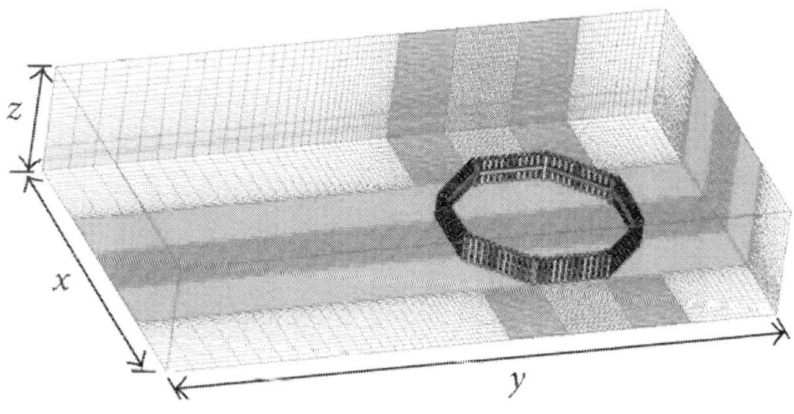

Figure 3: The analyzed domains of CFD.

The turbulence kinetic energy ,*k* , is obtained from the following equation [8, 9]:

$$\begin{aligned} &\frac{\partial}{\partial t}(\rho k) + \frac{\partial}{\partial x_i}(\rho k u_i) \\ &= \frac{\partial}{\partial x_j}\left[\left(\mu + \frac{\mu_t}{\sigma_k}\right)\frac{\partial k}{\partial x_j}\right] + P_k + P_b - \rho\varepsilon - Y_M + S_k. \end{aligned} \tag{2}$$

Whereas rate of dissipation ε can be obtained from the equation below:

$$\frac{\partial}{\partial t}(\rho\varepsilon)+\frac{\partial}{\partial x_i}(\rho k u_i)=\frac{\partial}{\partial x_j}\left[\left(\mu+\frac{\mu_t}{\sigma_k}\right)\frac{\partial k}{\partial x_j}\right]$$
$$+C_{1\varepsilon}\frac{\varepsilon}{k}(P_k+C_{3\varepsilon}P_b)-C_{2\varepsilon}\rho\frac{\varepsilon}{k}+S_\varepsilon. \quad (3)$$

The term in the above equation, P_k represents the generation of turbulence kinetic energy due to the mean velocity gradients. P_b is the generation of turbulence kinetic energy due to buoyancy, while represents the contribution of the fluctuating dilatation in compressible turbulence to the overall dissipation rate. The model constants $C_{1\varepsilon}, C_{2\varepsilon}, C_\mu, \sigma_k$, and ρ_k have been determined by Laundar and Saplding [10]:

Linear Windbreak Fences

Figure 4 shows the dimensions of the CFD model for the linear windbreak fences analysis. Figure 5shows the analysis distances (1H–20H) and heights (LEVEL1–LEVEL4) of the linear windbreak fences. The analysis range, aimed at examining wind velocity distribution in the back side of the fences, varies according to the height of the fences. Analysis was made up to 800 cm (20H), with the fence height at 40 cm and up to 1,600 cm (20H), with the fence height at 80 cm. The analysis heights of Level1, Level2, and Level3 were set at 0 cm, 10 cm, and 20 cm, regardless of the fence height. Analysis of Level4 was made at 40 cm (Level4-1), 60 cm (Level4-2), and 80 cm (Level4-3), according to the height of the fences. Two types of linear windbreak fence were tested: single (TYPE1) and double (TYPE2). Two heights of fence installation were tested: one with no gap and one with a 15 cm gap from the ground. To assess the linear windbreak fence effects with different wind velocities, we tested wind velocities of 3 m/s and 6 m/s.

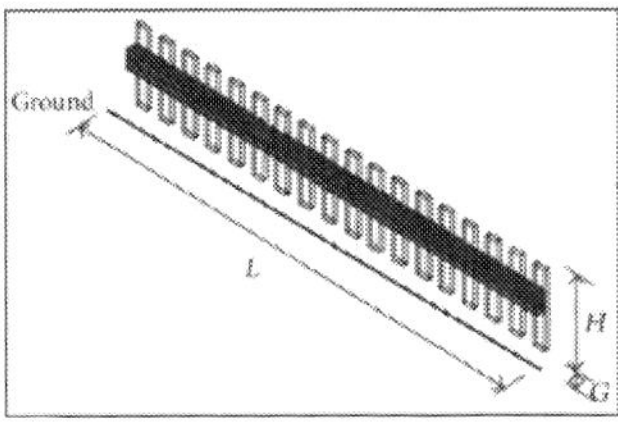

Single windbreak fence	
L	245 cm
H	40 cm, 60 cm, 80 cm
G	Gap 15 (ground + 15 cm) Gap 0 (ground 0 cm)

(a)

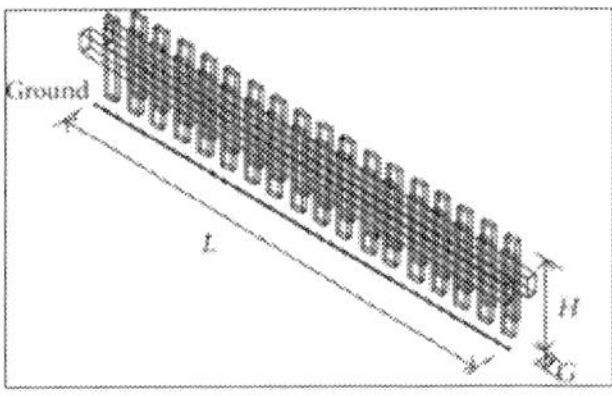

Double windbreak fence	
L	245 cm
H	40 cm, 60 cm, 80 cm
G	Gap 15 (ground + 15 cm) Gap 0 (ground 0 cm)

(b)

Figure 4: Dimension of the CFD Model for linear windbreak fences.

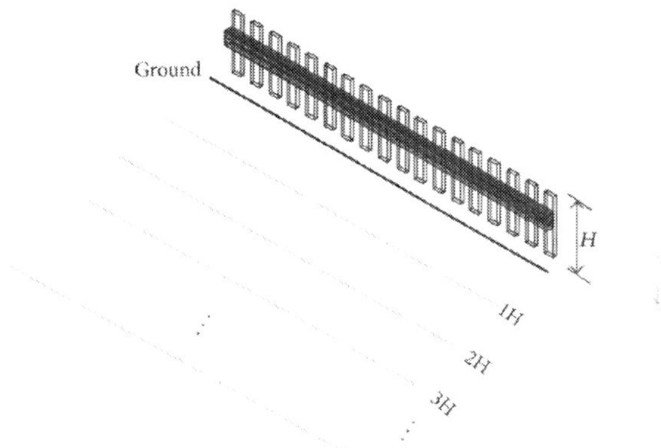

(a) Analysis distances

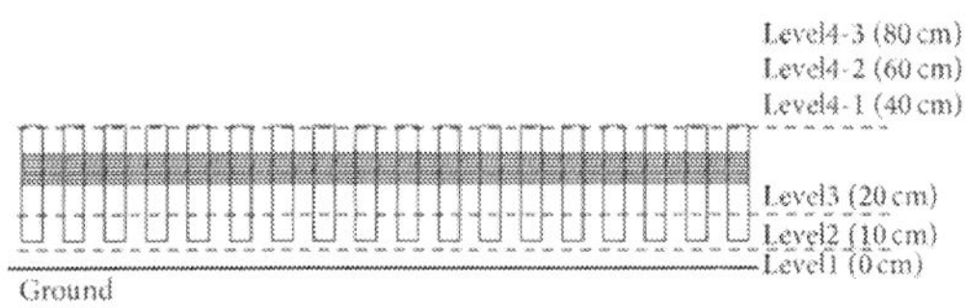

(b) Analysis height

Figure 5: Analysis distances (1H–20H) and heights (LEVEL1–LEVEL4) of the linear windbreak fences.

Simulation Results

This study has analyzed the downwind and vertical velocity distributions scaled by the height of the fence (H). It has interpreted the distance-based wind velocity distribution up to the range of 1H–20H. The vertical wind velocity distribution on the downwind side of the fences has been interpreted at the central point of the fence. Figures 6 and 7 show the wind velocity of linear windbreak fences that the distance is normalized with the height of the fence (H). The decrease in wind velocity on the downwind sides of the single fences and double fences (with the windbreak fence's vertical bars fixed on the ground) is influenced by the distance from the fences and the fence height.

Regardless of fence height, the wind velocity was regularly decreased to the power of 1H–7H. However, it was influenced by fence height when the distance was more than 7H. With a fence height of 40 cm, the wind velocity was almost regularly distributed up to 7H–20H on the downwind sides of the fences. However, with a fence height of 60 cm, the wind velocity was regularly distributed only up to 7H–15H and increased at heights higher than 15H. With a fence height of 80 cm, the wind velocity was regularly distributed only up to 7H–13H and increased at the heights higher than 13H. Unlike the single fences, the double fences showed great changes in wind velocity distant from their downwind sides. The wind velocity at the 1H point of the double fences decreased in comparison with the same point of the single fences by a maximum of 80% or more. Moreover, the wind velocity at the positions of 3H and 6H, distant from the downwind side of the fence, rapidly decreased to 0.1 m/s regardless of experimental wind velocity.

The wind velocity increased regardless of fence height at positions more distant than 6H. These phenomena, different from those with single fences, are believed to occur due to the thickness of the fences increasing by double. The wind velocity distribution at the vertical heights of 0–20 cm (Level1-Level3) of the double fences was almost identical up to 1H–3H. However, within the range of 3H–7H, the wind velocity varied with fence heights. While the wind velocity was the lowest at the 20 cm (Level3) point of the vertical height of the fences, it remained constant at the 80 cm point (Level4-3), regardless of vertical fence heights. Figure8 shows the analysis at the fence with a 60 cm (Level4-2) height, in order to identify the effects of the gap between the vertical bars of the windbreak fence and the ground. In the case of the single fences, the fastest wind velocity was measured at the point of 1H, distant from the downwind side of the fences at the vertical heights of 0 cm (Level1) and 10 cm (Level2), but the wind velocity decreased until the distance reached the 7H point, after which the wind velocity was almost identically maintained. The wind velocity at the vertical heights of 20 cm (Level3) and 60 cm (Level4-2) was almost regularly distributed at every distance for interpretation, but the wind velocity distribution was 35% slower at 1H–7H, with vertical heights of 0 cm (Level1) and 10 cm (Level2). This was because of gaps in the vertical windbreak fence bars, at a height of 15 cm from the ground. In distance-based changes in wind velocity, the effects of a decrease in wind velocity were less than in the cases of the windbreak fence's vertical bars set on the ground. In the case of the double fences, like the case of the single fences, the wind velocity decreased at the points of 0 cm (Level1) and 10 cm (Level1), when the distance was within the range from 1H to 7H, and then increased when the distance was more than 7H.

The wind velocity decreased by 83% or more at the distance of 7H than at 1H. However, with fence heights of 20 cm (Level3) and 60 cm (Level4-2), the wind velocity continuously increased after the greatest amount decreased around 1H. Figure 9 shows wind velocity flow inside of the double windbreak fences by height. The reference wind velocity was 3 m/s, and the wind velocity flow on the downwind side showed up diversely, according to fence heights. The effects of reduced wind velocity on the downwind side of the fences can be identified in a wide area. It was identified that wind velocity greatly decreased, as flow separated from the upper part of the fence within a certain distance range was reattached on the bottom. Wind velocity decrease rates were great around 8H, at a fence height of 40 cm, and around 2H–4H at a fence height between 60 cm and 80 cm.

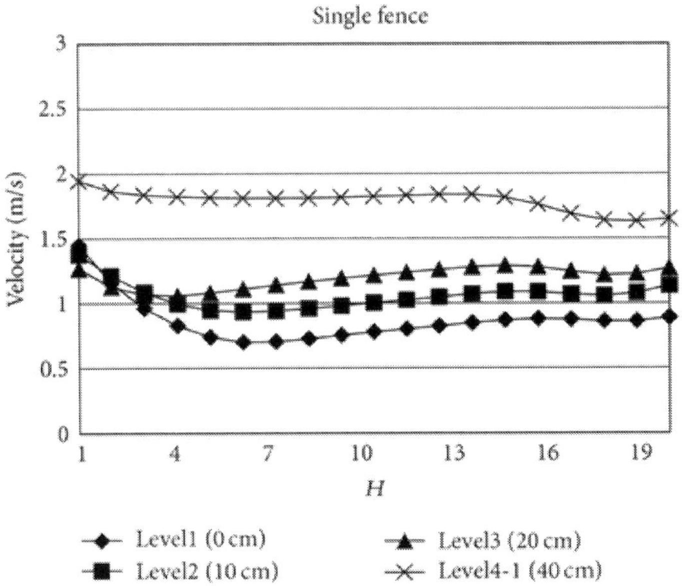

(a) TYPE1 (height of fence: 40 cm)

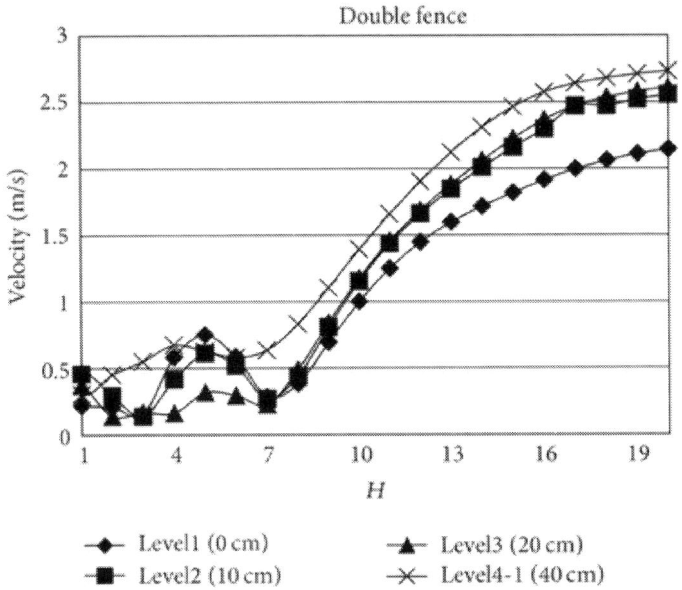

(b) TYPE2 (height of fence: 40 cm)

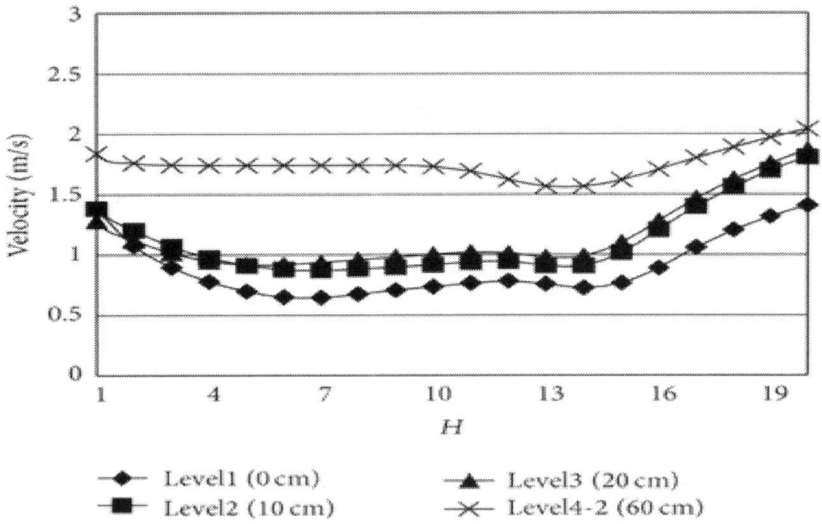

(c) TYPE1 (height of fence: 60 cm)

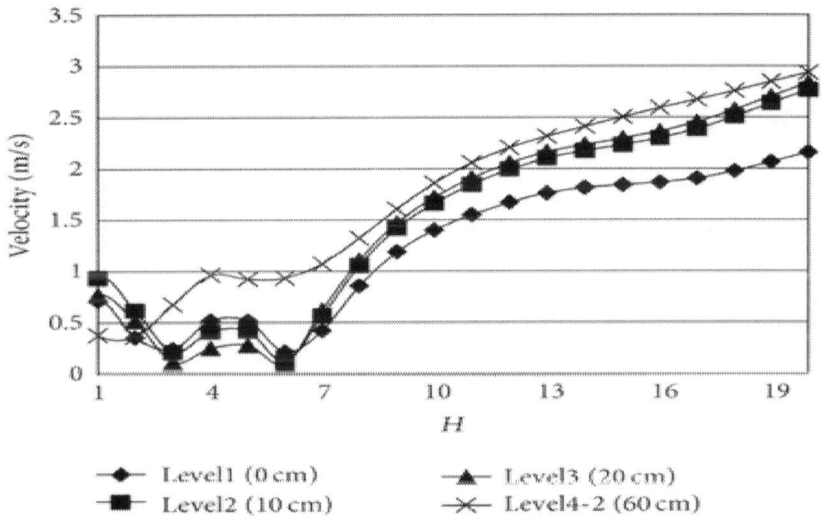

(d) TYPE2 (height of fence: 60 cm)

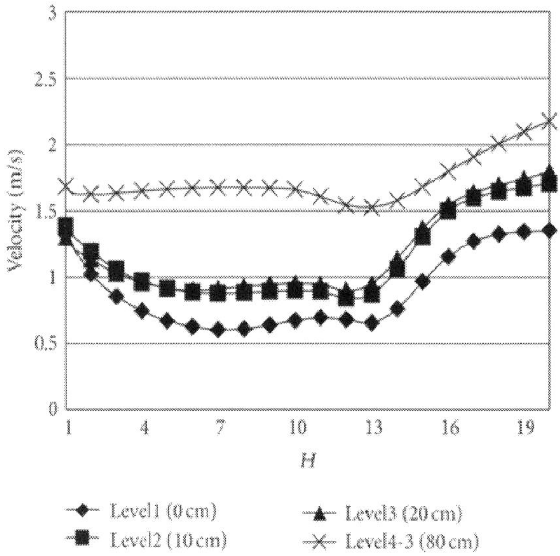

(e) TYPE1 (height of fence: 80 cm)

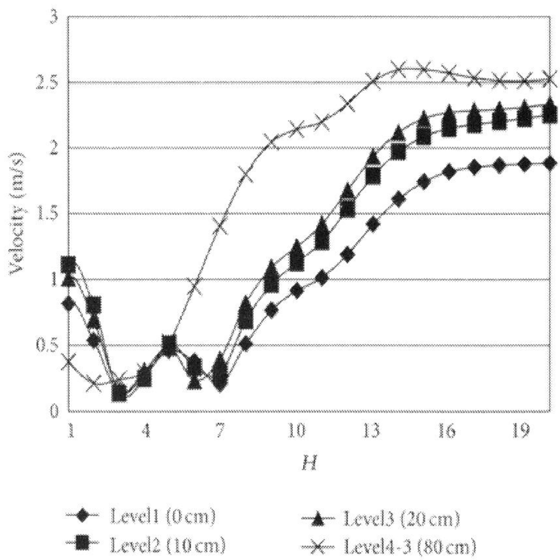

(f) TYPE2 (height of fence: 80 cm)

Figure 6: Wind velocity distribution of the linear windbreak fence by distance (GAP 0 cm, reference velocity = 3 m/s).

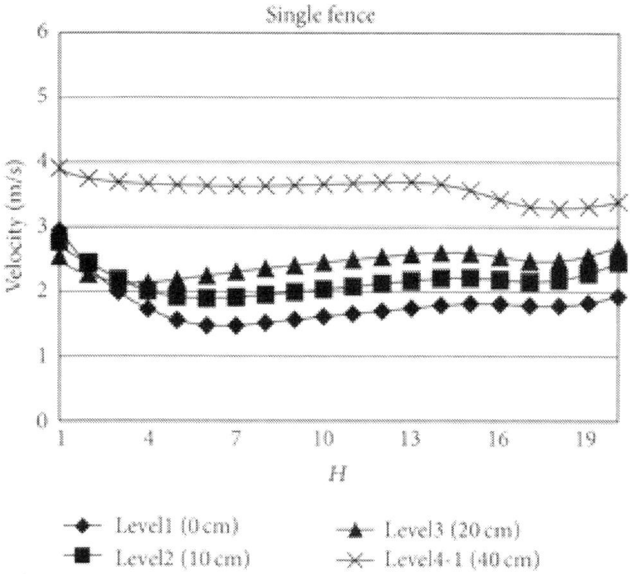

(a) TYPE1 (height of fence: 40 cm)

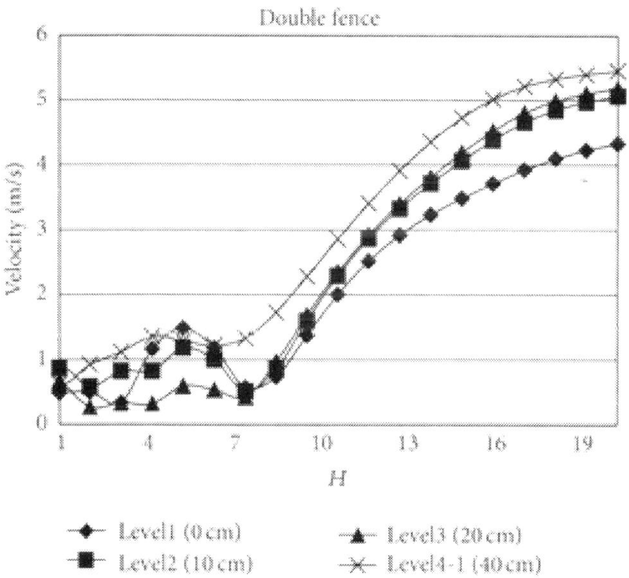

(b) TYPE2 (height of fence: 40 cm)

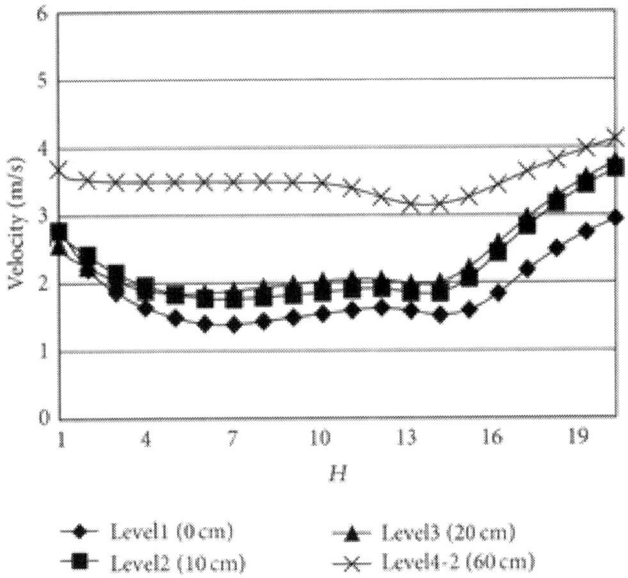

(c) TYPE1 (height of fence: 60 cm)

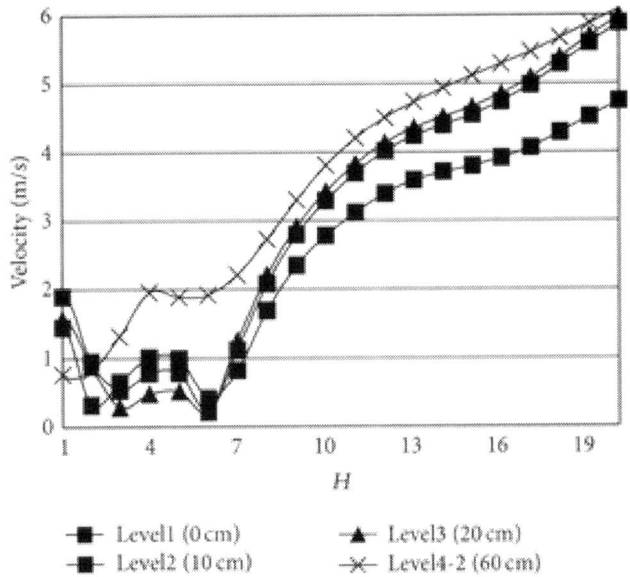

(d) TYPE2 (height of fence: 60 cm)

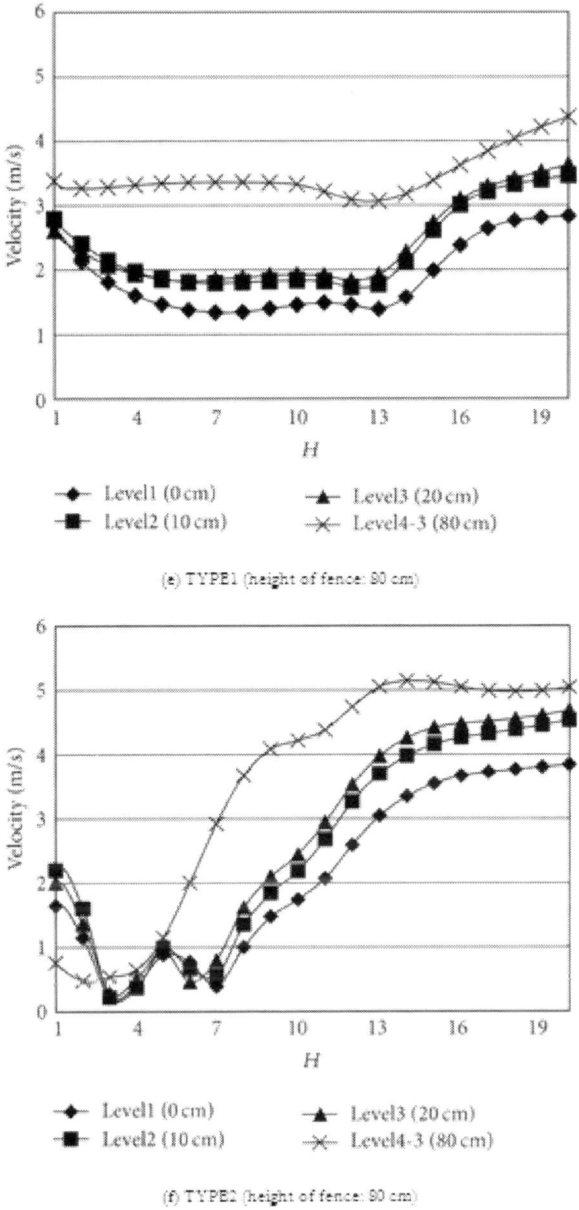

Figure 7: Wind velocity distribution of the linear windbreak fence by distance (Gap 0 cm, reference velocity = 6 m/s).

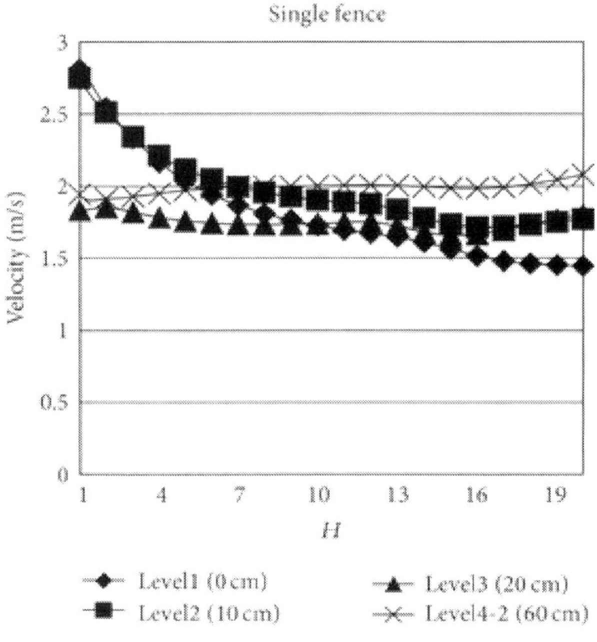

(a) TYPE1: height 60 cm, velocity 3 m/s

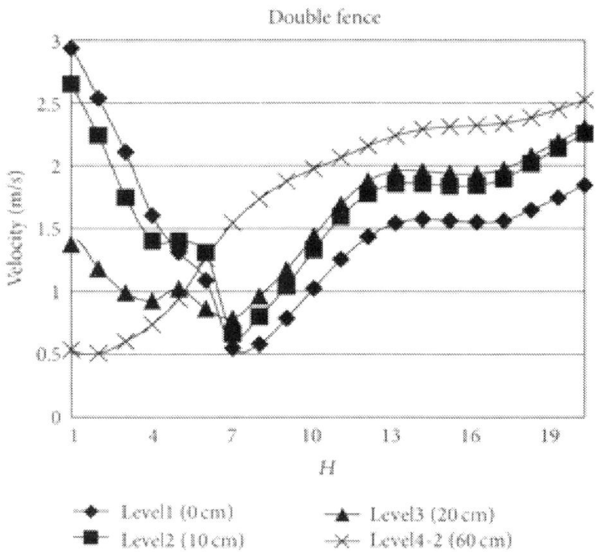

(b) TYPE2: height 60 cm, velocity 3 m/s

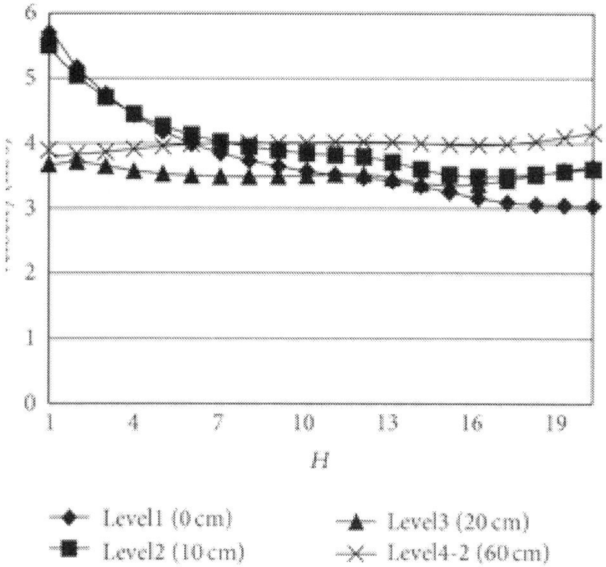

(c) TYPE1: height 60 cm, velocity 6 m/s

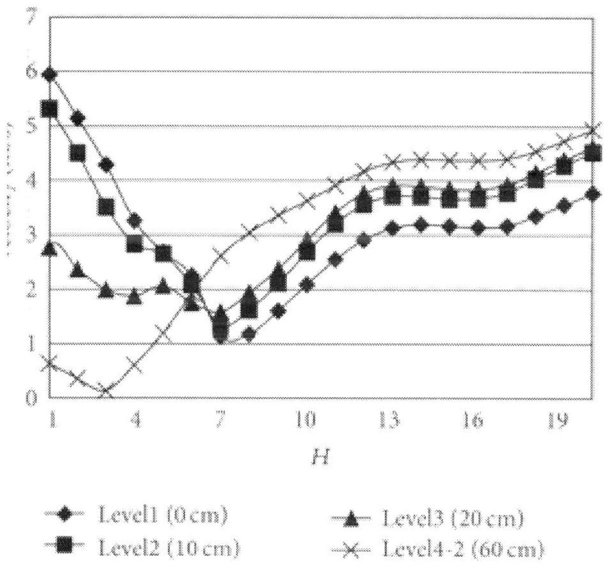

(d) TYPE2: height 60 cm, velocity 6 m/s

Figure 8: Wind velocity distribution of the linear windbreak fence by distance (Gap 15 cm, reference velocity = 3 m/s, 6 m/s).

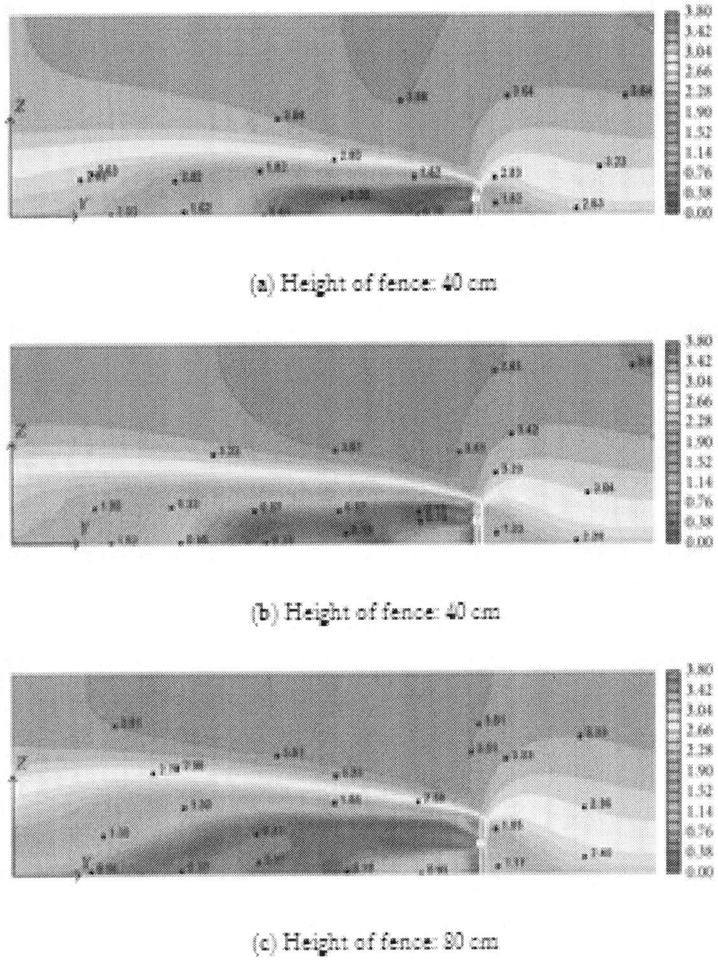

(a) Height of fence: 40 cm

(b) Height of fence: 40 cm

(c) Height of fence: 80 cm

Figure 9: Wind velocity flow the double windbreak fence by height (velocity 3 m/s).

Octagonal Windbreak Fence

Based on a wind flow analysis of the linear windbreak fence, we found a double fence with a 6 m length that was fixed to the ground to be the most effective. The wind direction will vary in a natural environment, so we analyzed the internal wind flow characteristics by installing an octagonal type windbreak, while considering the direction of the wind. After the octagonal windbreak fence was installed, we determined the width of the fence. Two fence widths (4 m and 6 m) were tested, based on the simulation of the linear fence. Figure 10 shows a CAD (Computer Aided Design) drawing of the fence. The baseline wind velocities were 3 m/s and 6 m/s at fence height. The wind direction was from the front.

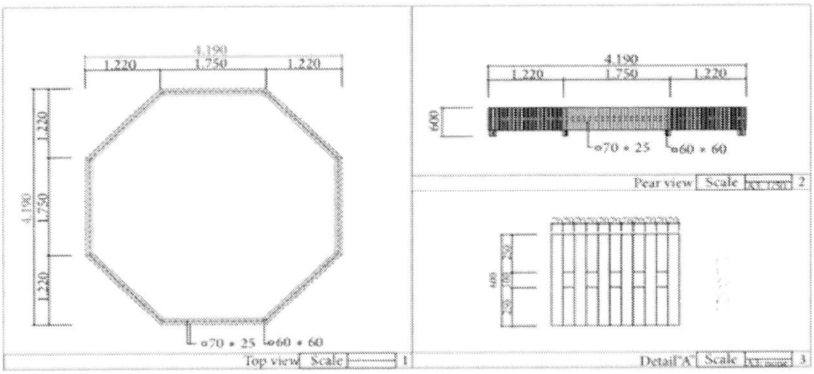

(a) CASE 1 (width: 4 m)

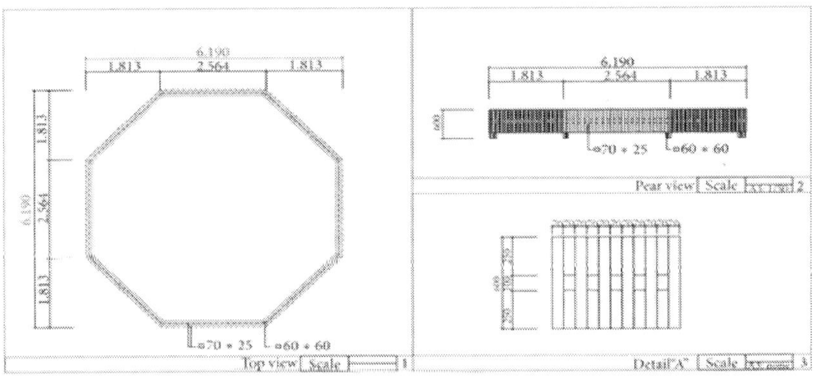

(b) CASE 2 (width: 6 m)

Figure 10: Specifications of the octagonal windbreak fence.

Simulation Results

Figures 11 and 12 show the wind velocity flow inside and outside the fence, at heights of 10 cm and 20 cm, depending on the width of the octagonal windbreak fence. The internal wind flow appeared to be constant, regardless of the width. There were large wind velocity decrease effects at the left and right edges of the vertical windbreak fence and in the center of the fence. The wind velocity decrease in the center of the fence had a "U" shape. The 6 m fence had a more obvious "U"-shaped wind decrease than the 4 m fence. The wind flow distribution was not significantly different as the wind velocity increased. The magnitude of the wind velocity decrease in the center of the fence's inside was larger with the 4 m fence than it was with the 6 m fence. The wind flow outside the windbreak fence was almost the same, so it appeared that the external wind flow was not affected by the width of the fence. Figure 13 shows the wind velocity flow inside the fence by height. When the fence was detached from the ground by 15 cm, the wind entering below the bottom of the fence affected the wind velocity at the depth on the opposite side of the fence and affected the internal wind flow. However, when the windbreak fence's vertical bars were attached to the ground, the wind only affected up to a certain distance and it had no major effects on the internal wind flow. The vortex shedding, separated from the top of the upwind side of the fence, had effects on the opposite side. The vortex created at the upwind of the 4 m fence crossed to the opposite side of the fence but did not cross to the opposite side of the 6 m fence. Instead, it remained inside the fence and affected the internal wind flow.

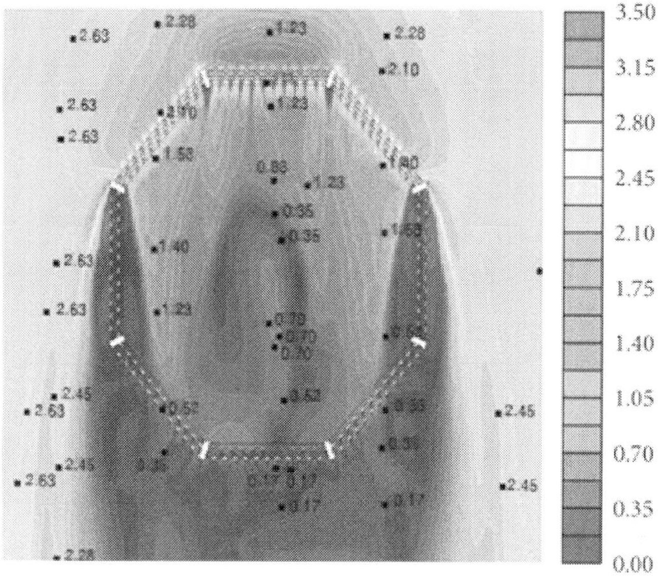

(a) Width of windbreak fence: 4 m (velocity 3 m/s)

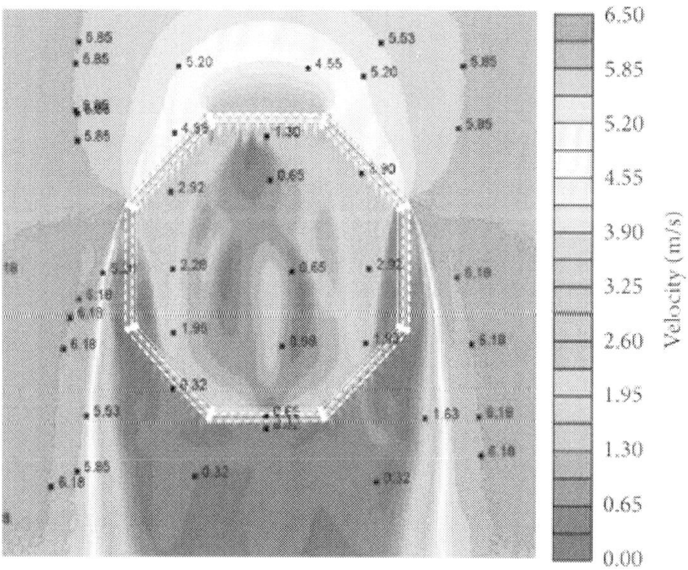

(b) Width of windbreak fence: 4 m (velocity 6 m/s)

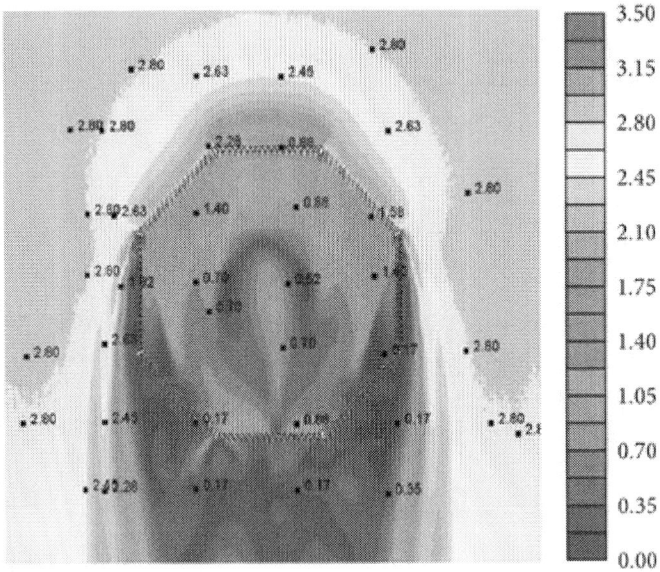

(c) Width of windbreak fence: 6 m (velocity 3 m/s)

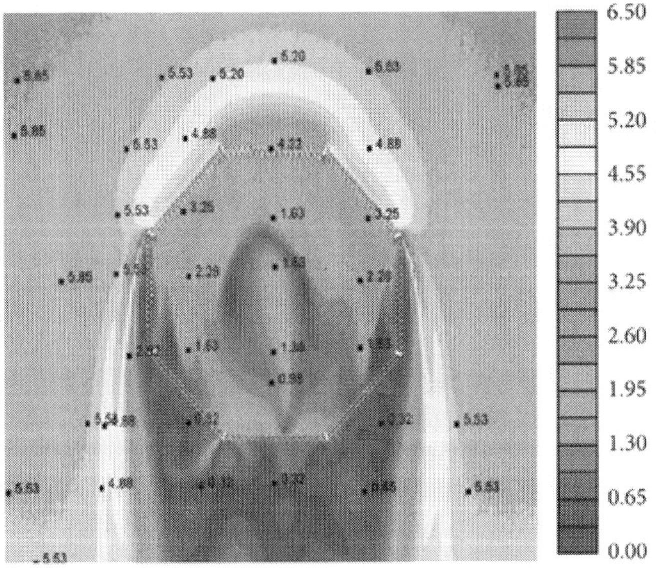

(d) Width of windbreak fence: 6 m (velocity 6 m/s)

Figure 11: Internal wind velocity flow at a measurement height of 10 cm according to the windbreak fence width.

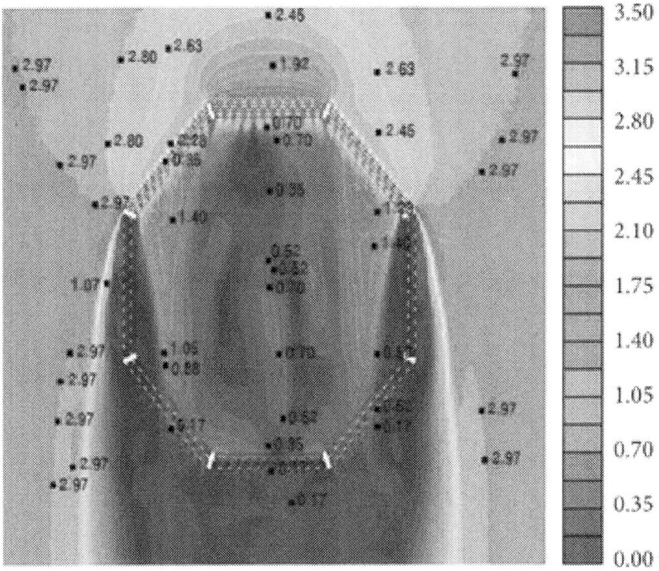

(a) Width of windbreak fence: 4 m (velocity 3 m/s)

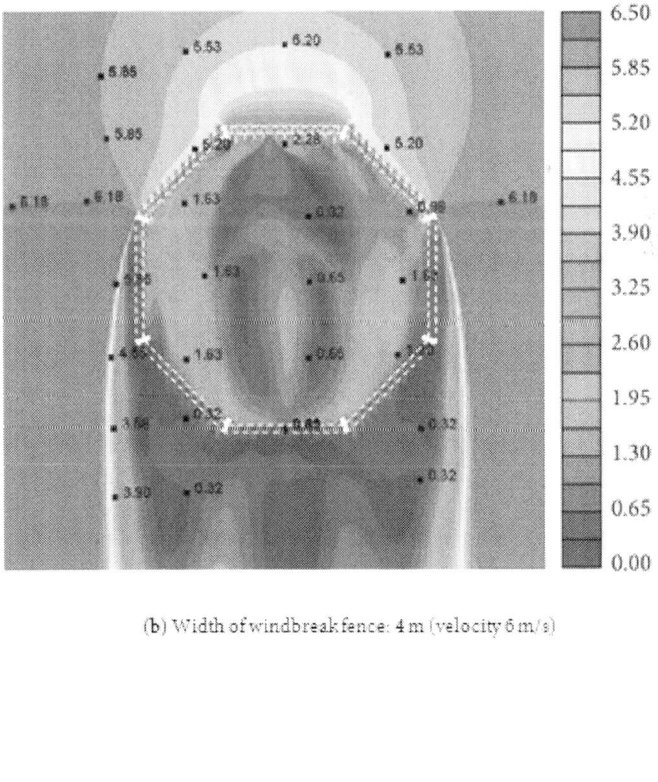

(b) Width of windbreak fence: 4 m (velocity 6 m/s)

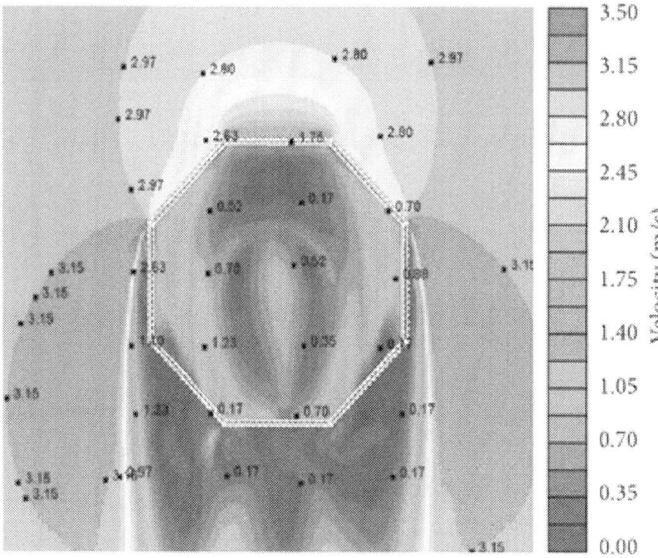

(c) Width of windbreak fence: 6 m (velocity 3 m/s)

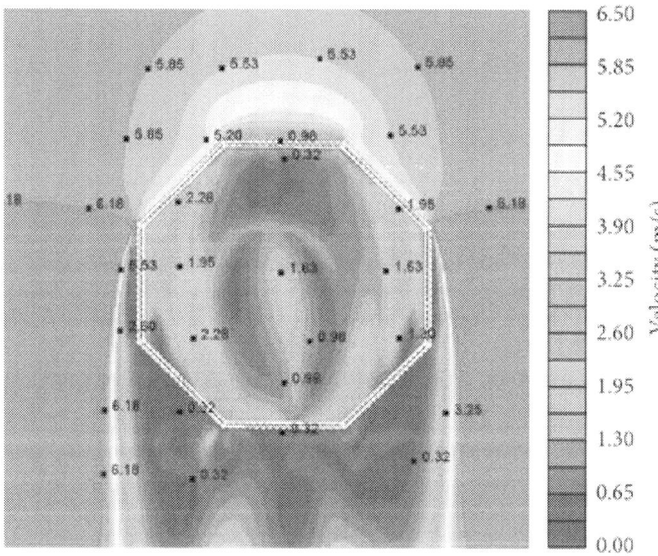

(d) Width of windbreak fence: 6 m (velocity 6 m/s)

Figure 12: Internal wind velocity flow at a measurement height of 20 cm according to the windbreak fence width.

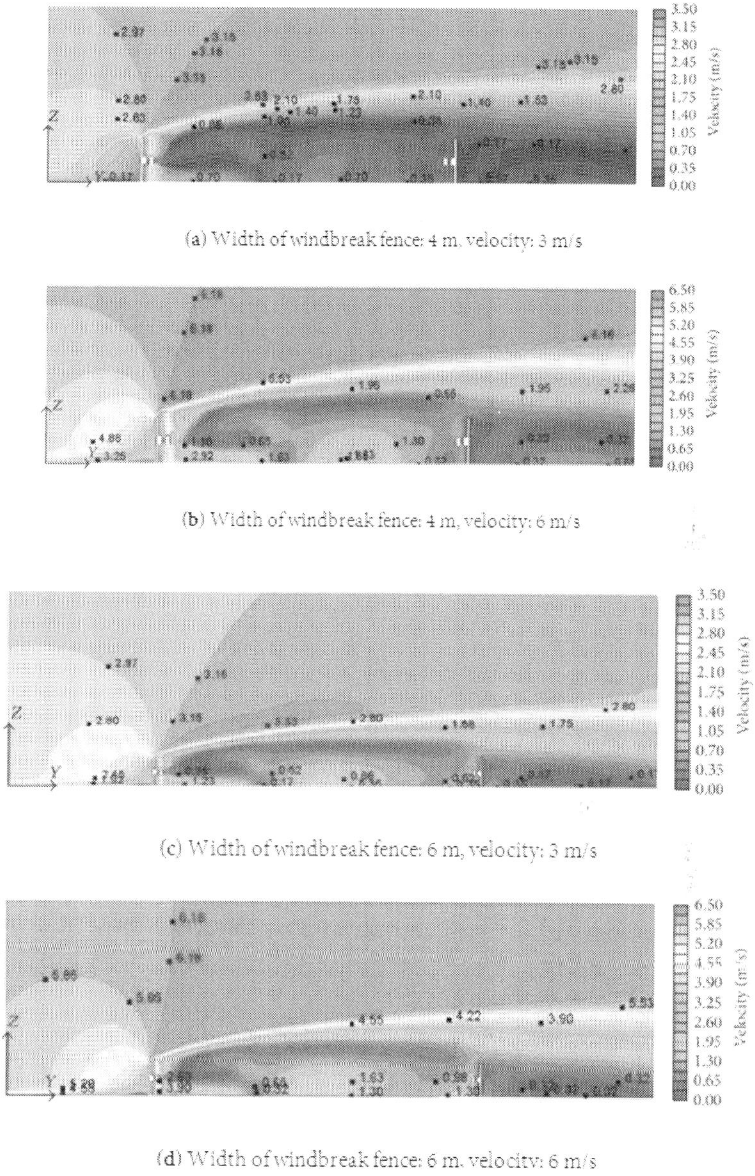

Figure 13: Wind velocity flow inside the fence by height.

Wind Tunnel Test

A wind tunnel test was performed to analyze the wind flow inside the windbreak fence. Wind tunnel tests were conducted in an open-type wind tunnel located at the Department of Architectural Engineering, Chonbuk National University, which has a test section of 1.5 m (width) × 1.7 m (height) and 20 m (length). The boundary layer flow condition, representing natural wind flow over suburban terrain, indicated that the power law exponent of the mean longitudinal wind velocity profile was 0.15 and that the longitudinal turbulence intensity was about 16% at the top of the building model. Terrain category C, as described in the Standard Design Loads for Building, 2010, Architectural Institute of Korea, was simulated over the test section using spires and wooden blocks [11, 12]. The vertical distribution of the mean longitudinal wind velocities and the longitudinal turbulence intensities is shown in Figure 14. The wind tunnel test model was made of balsa wood. Table 1 shows the full-scale size of the experimental model and the model size. Two windbreak fence widths (6 m and 4 m) were used in the experiment. The windbreak fence was attached to the ground and the scale of the model was 1/5. The baseline wind velocity was assumed to be 3 m/s or 6 m/s at the height of the windbreak fence. Because the model was reduced, the wind velocity had to be decreased accordingly, which was calculated using (6). The scale of the wind velocity was calculated as the square root of the geometrical scale of the model. We used 1.34 m/s and 2.68 m/s as the experimental wind velocities. The heights inside the fence model were 2 cm and 4 cm, which were equivalent to full-scale sizes of 10 cm and 20 cm. Figures 15 and16 show the wind velocity measurement locations for the windbreak fences, measuring 0.8 m wide (full-scale size: 4 m, CASE 1) and 1.2 m wide (full-scale size: 6 m, CASE 2), respectively. We measured 25 points for CASE 1 and 45 points for CASE 2. The width of the fence was constant because it was an octagonal, but the internal wind flow of the 6 m width was measured using a rectangular shape. The full-scale wind flow inside the fence was analyzed by measuring (horizontally and vertically) inside the fence at regular intervals:

$$\frac{V_{\text{model}}}{V_{\text{full}}} = \sqrt{\frac{B_{\text{Model}}}{B_{\text{full}}}}. \tag{6}$$

Table 1: Specifications of the windbreak fence used in the wind tunnel test (unit = m).

	Width		Height	
	Full-scale	Model	Full-scale	Model
Case A	4	0.8	0.6	0.12
Case B	6	1.2	0.6	0.12

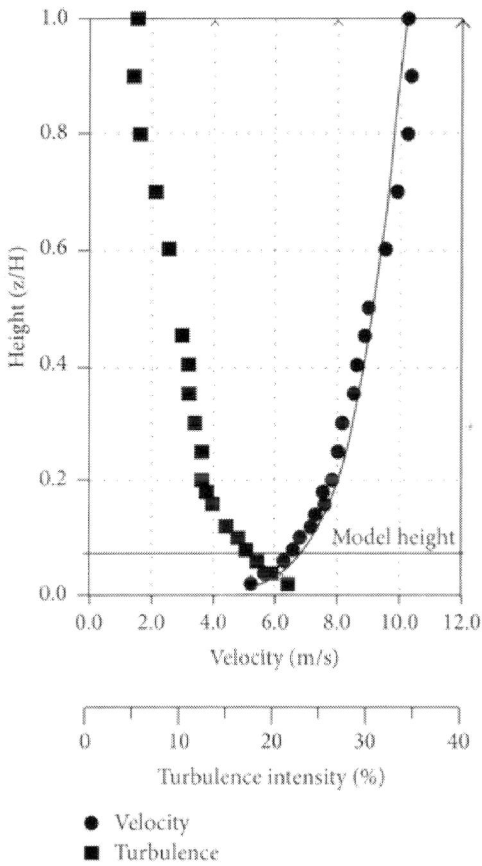

Figure 14: Vertical distribution of the mean longitudinal wind velocities and turbulence intensities.

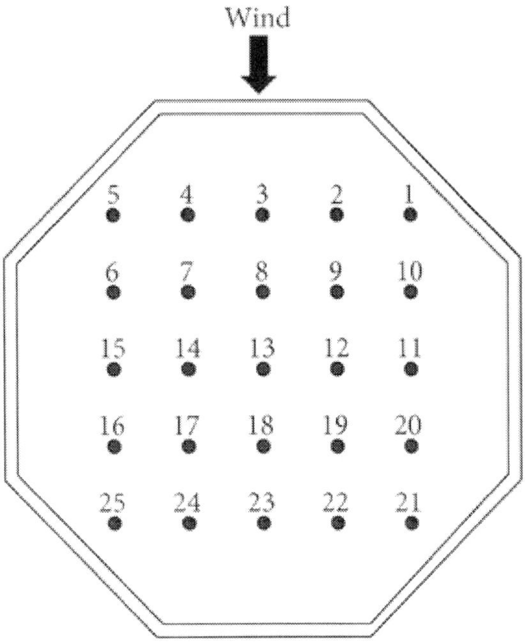

(a) Velocity measurement location

(b) Model

Figure 15: Wind velocity measurement locations and CASE 1 model installed in the wind tunnel.

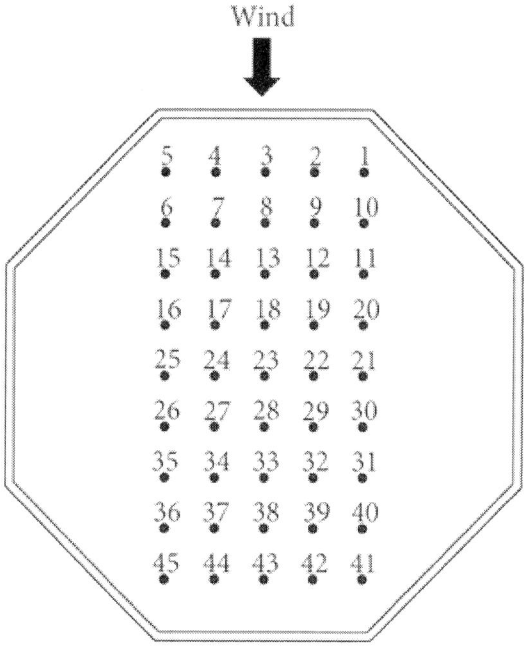

(a) Velocity measurement location

(b) Model

Figure 16: Wind velocity measurement locations and CASE 2 model installed in the wind tunnel.

The average wind velocity and turbulence intensity changed according to the location analyzed. The non-dimensional average velocity used in the experimental analysis was calculated using (7), while the non-dimensional turbulence intensity was calculated using

$$\frac{V}{V_0}, \quad (7)$$

where V: average wind velocity in the measurement location and : wind velocity at the height of the fence

$$\frac{I}{I_0}, \quad (8)$$

where I : turbulence intensity in the measurement location and I_0: turbulence intensity at the height of the fence.

Figures 17 and 18 show the distribution of wind velocity change rates within the windbreaks, whose widths are 4 m and 6 m. In the case of the fences with a width of 4 m, the wind velocity decrease rate was greatest at the center of the fences. When the measured height was 2 cm, the wind velocity decrease rate was greater in the downwind part of the fences, where wind blows, than in the upwind part. When the measured height was 4 cm, the rates of wind velocity decreases in the downwind and upwind parts of the fences were the same. This is thought to be due to the fact that the wind separated at the upwind parts of the fences hit the downwind parts of the fences and then flew downward. Even though the reference wind velocity increased, the wind velocity decrease rates at the center of the fences were the same, at 78%. The wind velocity decrease rate for the internal wind flow was the same, according to the wind velocity increase with the measurement height. With the width of 6 m, the wind flow was measured in the shape of a rectangle and the wind flow was represented as a rectangular shape. In contrast to the width of 4 m, the wind velocity decrease rate was the highest at the upwind part of the width of 6 m from where the wind blew. The wind flow characteristics, according to the height and baseline wind velocity, were the same as those with the width of 4 m. However, the increased fence width increased the wind velocity at the back of the fence. Figures 19 and 20 show the distribution of the turbulence intensity inside the windbreak fence, according to the fence width. The turbulence intensity inside the windbreak fence was lowest at the upwind part and at the edge of the fence. Above 0.5 H, there was a constant turbulence intensity distribution. As the wind velocity increased, the turbulence intensity became smaller and it was constant, regardless of the height. This confirms that the turbulence was constant inside the windbreak fence. The width of 6 m (CASE 2) had a lower turbulence intensity than the width of 4 m (CASE 1).

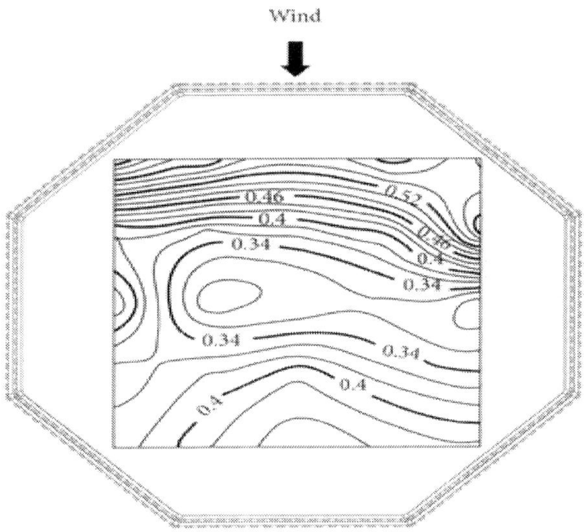

(a) Height of measurement: 2 cm, velocity: 1.38 m/s

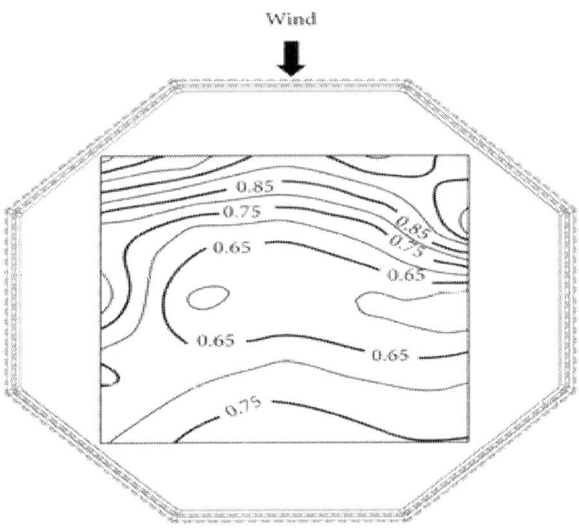

(b) Height of measurement: 2 cm, velocity: 2.68 m/s

(c) Height of measurement: 4 cm, velocity: 1.38 m/s

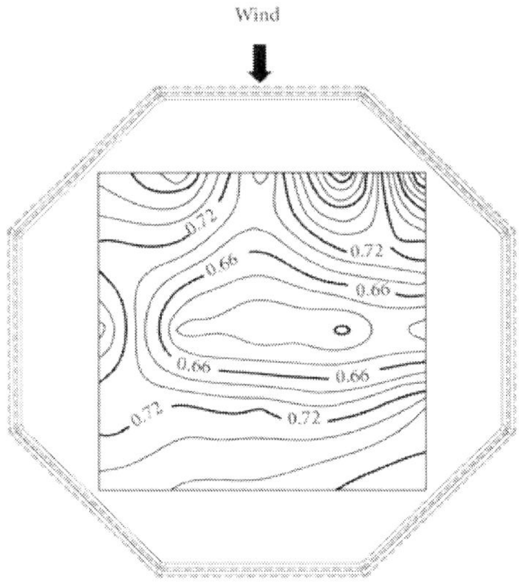

(d) Height of measurement: 4 cm, velocity: 2.68 m/s

Figure 17: Distribution of wind velocity within the windbreak (CASE 1).

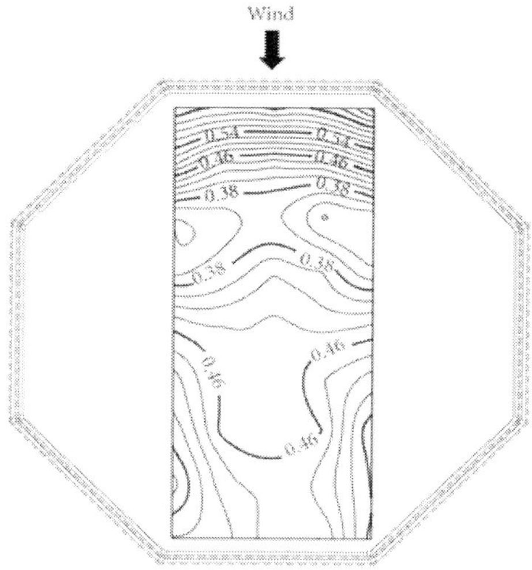

(a) Height of measurement: 2 cm, velocity: 1.39 m/s

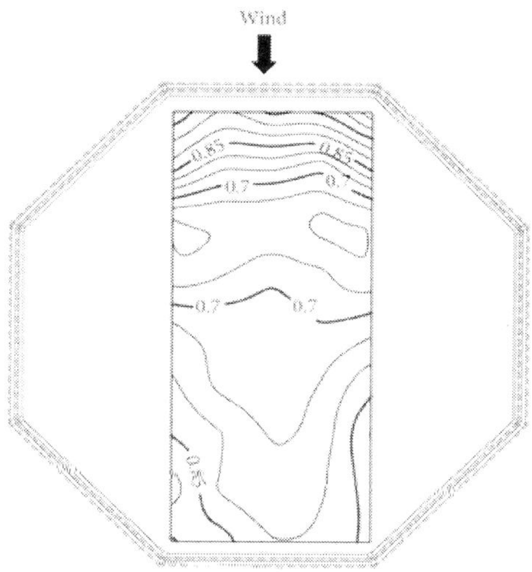

(b) Height of measurement: 2 cm, velocity: 2.69 m/s

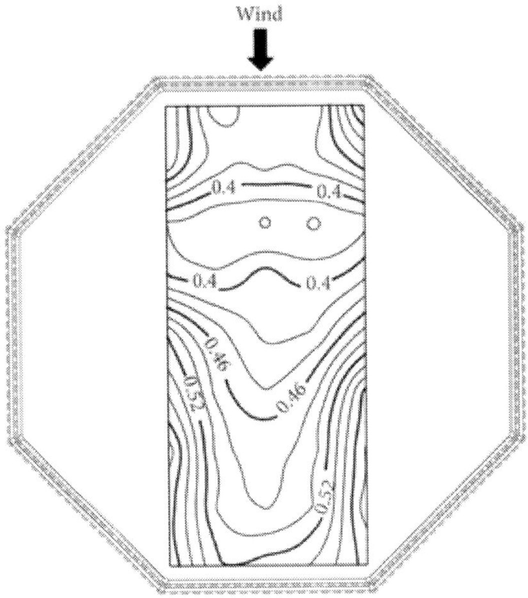

(c) Height of measurement: 4 cm, velocity: 1.38 m/s

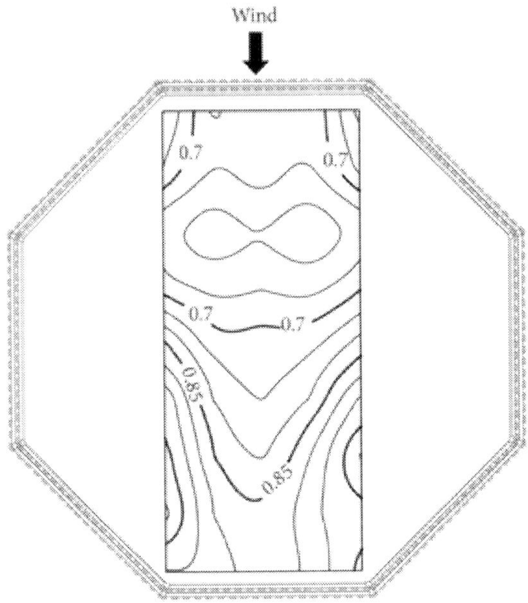

(d) Height of measurement: 4 cm, velocity: 2.68 m/s

Figure 18: Distribution of wind velocity within the windbreak (CASE 2).

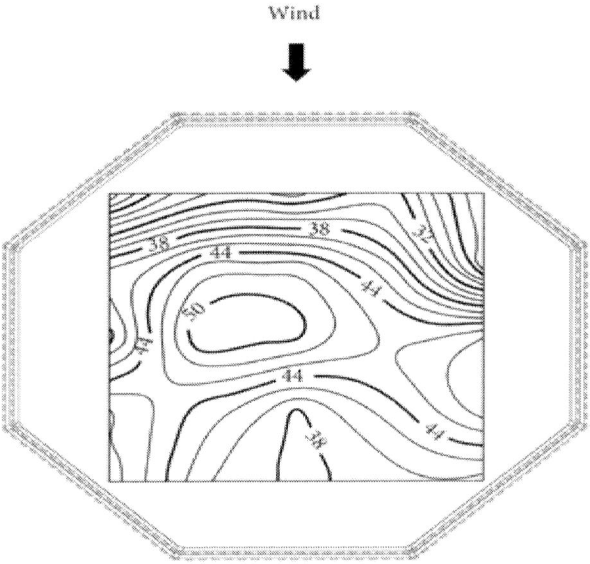

(a) Height of measurement: 2 cm, velocity: 1.38 m/s

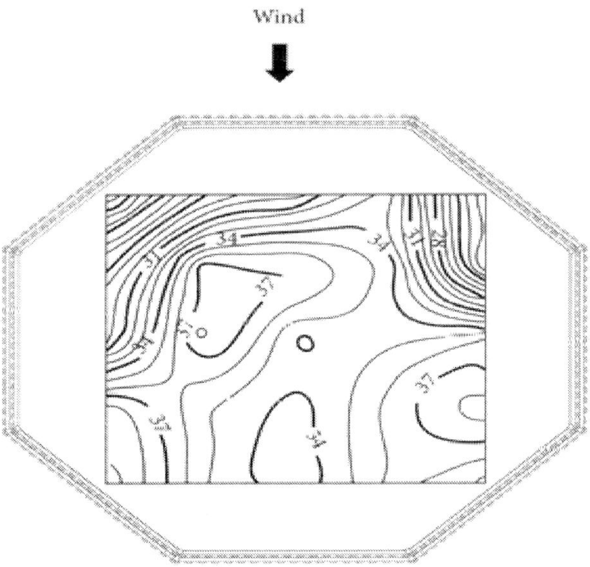

(b) Height of measurement: 2 cm, velocity: 2.68 m/s

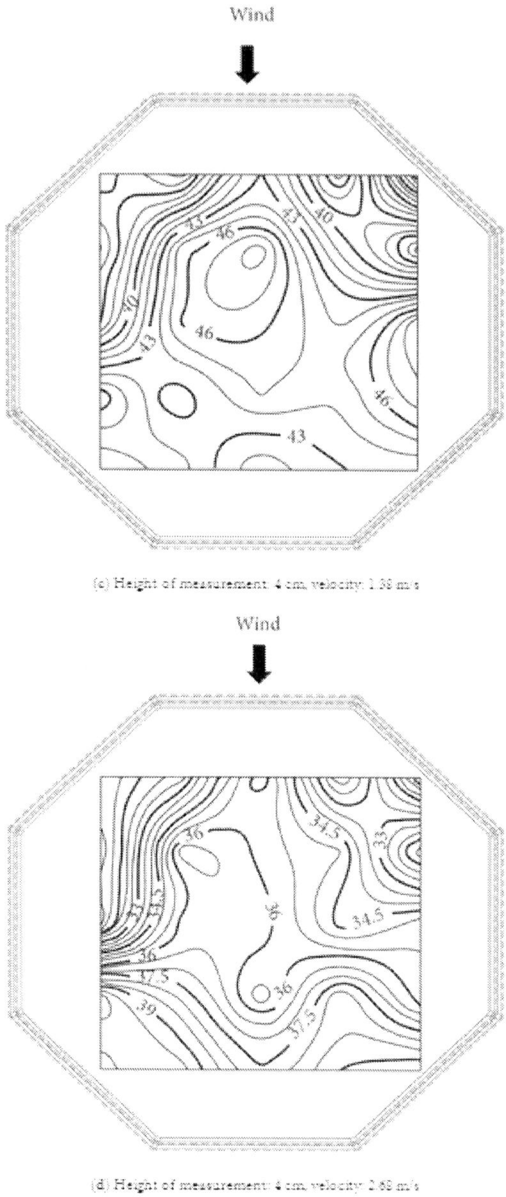

Figure 19: Distribution of the turbulence intensity inside the windbreak fence. (CASE 1).

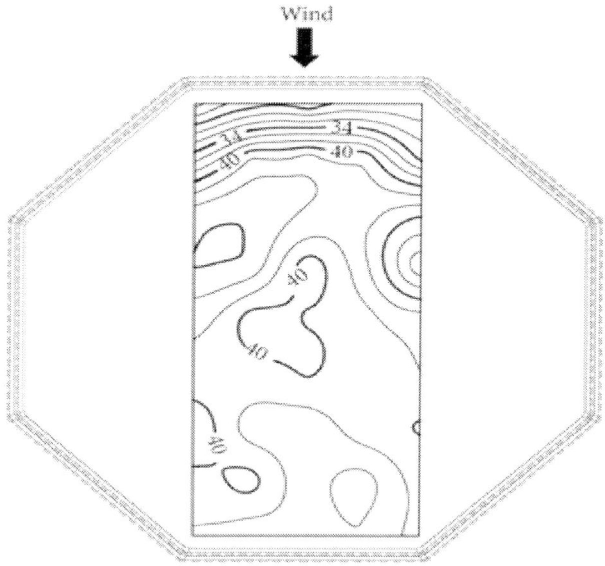

(a) Height of measurement: 2 cm, velocity: 1.38 m/s

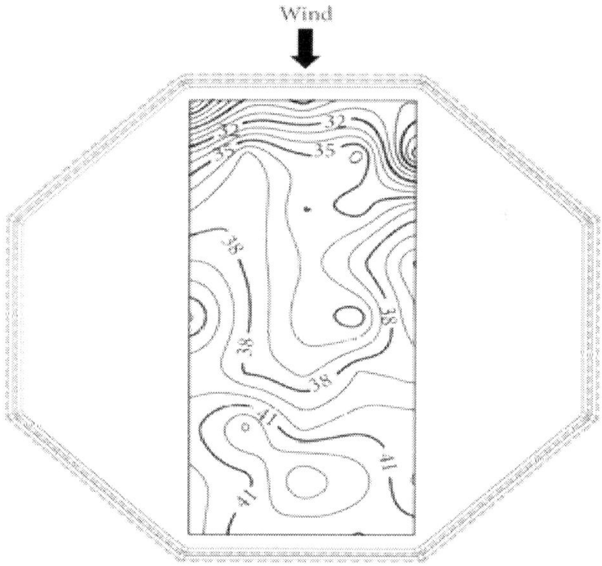

(b) Height of measurement: 2 cm, velocity: 2.68 m/s

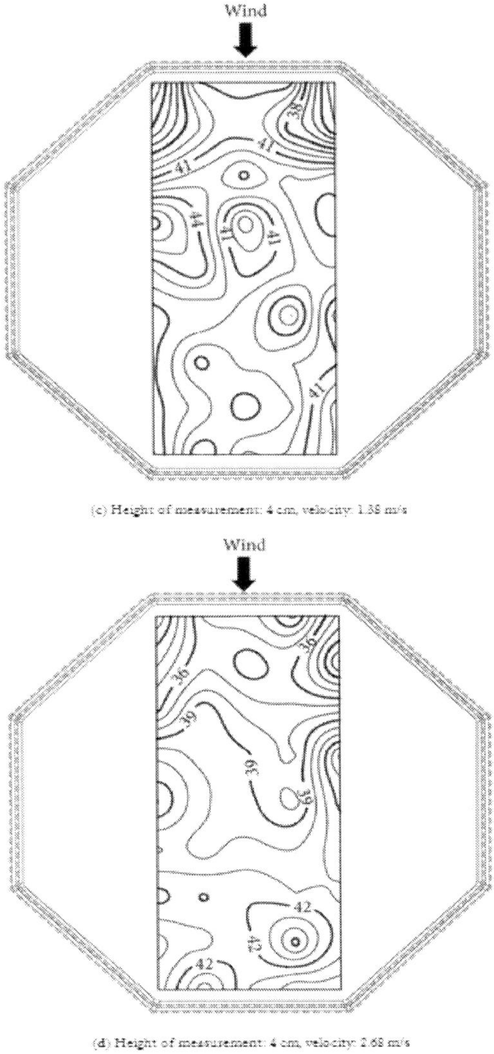

Figure 20: Distribution of the turbulence intensity inside the windbreak fence (CASE 2).

Comparative Analysis of the CFD and Wind Tunnel Tests

We compared the wind tunnel test results and the CFD results. The CFD simulation was performed using the same size of model and the same wind velocity as the wind tunnel test. The CFD simulation and wind tunnel test results were compared with CASE 1 (fence width 4 m). We used 1.38 m/s and 2.68 m/s as the wind velocities in the wind tunnel test. The wind velocity distribution heights inside the windbreak fence were 2 cm and 4 cm. Figure 21 shows the wind velocity distribution and the correlation coefficients for the wind tunnel test and CFD, according to the wind velocity. The correlation coefficients of the CFD and wind tunnel tests were distributed in a range of 0.7-0.8, regardless of wind velocity and measurement height. The correlation coefficient was close to 1, which confirmed that there was a strong correlation between the wind tunnel test and the CFD results.

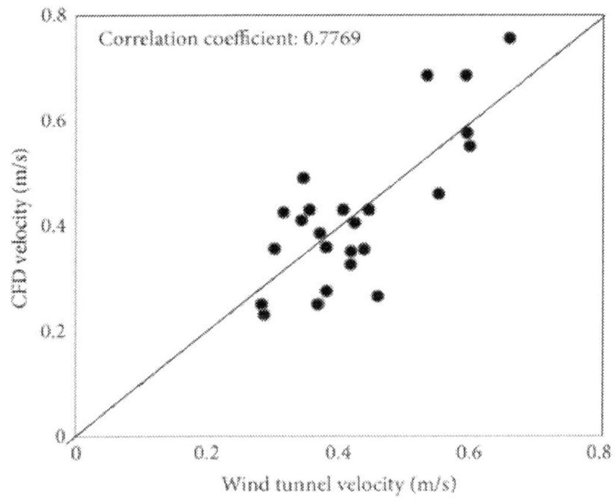

(a) Width: 4 m, velocity: 3 m/s, height: 2 cm

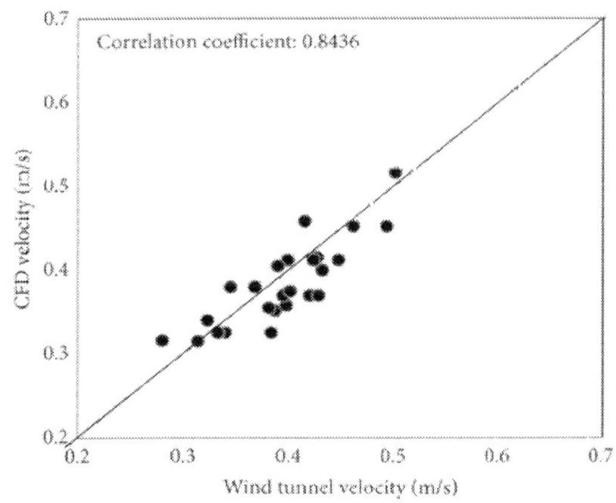

(b) Width: 4 m, velocity: 3 m/s, height: 4 cm

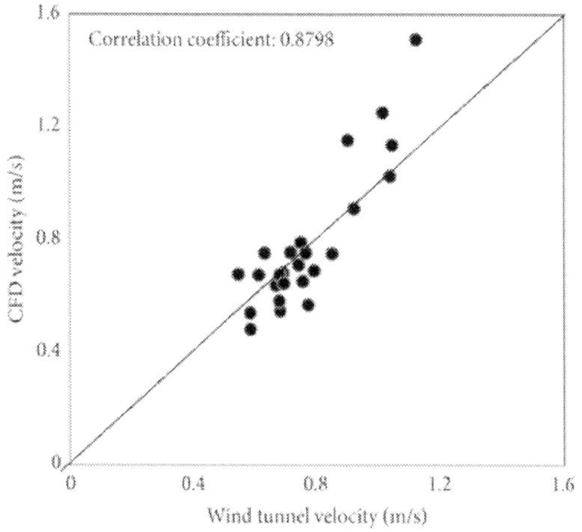

(c) Width: 4 m, velocity: 6 m/s, height: 2 cm

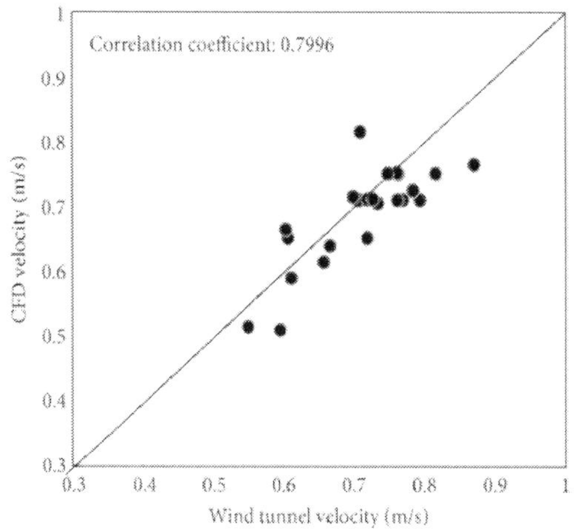

(d) Width: 4 m, velocity: 6 m/s, height: 4 cm

Figure 21: Wind velocity distribution and correlation coefficients for wind tunnel test and CFD.

Field Measurements

The windbreak fences for field measurement included two fences, one with a width of 6 m and one with a width of 4 m. The height of the windbreak fences was constant at 60 cm. The windbreak's vertical bars were fixed on the bottom. Table 2 shows the components of the models used for field measurement. The measuring instrument used for field measurement consisted of four three-cup anemometers (JY-WS161C) and two data loggers (JDL-74A) to automatically store measured data. A total of four anemometers were installed, one within each of the three windbreak models and one outside the windbreak models, between models A and C.

The anemometers installed at the center inside the windbreak fences had a height of 20 cm. Figure 22 shows the arrangement plan and measurement process of anemometers. The anemometers had received a performance test by the Korea Meteorological Industry Promotion Agency and were verified to have good performance. The period of field measurement was 26 days, from November 4th to November 29, 2012. The anemometers were installed in the Gochang weather station (35°20'N, 126°35'E). Measurement of average and maximum wind velocities was taken over 24 hours, at an interval of one hour. Figure 23 shows average and maximum wind velocities during each time. As the wind velocity was marked more quickly on the anemometers installed outside the models, it is judged that there were wind velocity decrease effects within the windbreaks. The wind velocity decrease rates within the windbreaks varied with the windbreaks' widths.

The average wind velocity decreased by 27%–63% inside the windbreak fences, with respect to the velocity outside. The maximal wind velocity decreased by 30%–50% inside the windbreak fences, with respect to the velocity outside. The windbreak fence width of 4 m (model B) turned out to be more effective in decreasing wind velocity rates than the width of 6 m (model A, model C). Using the windbreak width of 6 m, this study has compared the wind velocities within the fences, in the case of the windbreak's fixed vertical bars on the ground and separated from the ground. The maximal velocity was identical in model A, where the windbreak vertical bars were fixed on the ground, and in model C, where they were separated from the ground. However, the average wind velocity turned out to decrease by 10% or more at 9–19 hours in Model A, where the windbreak vertical bars were fixed on the ground, with respect to Model C. The fixed, vertical windbreak bars on the ground were more effective in achieving a significant reduction in wind velocity. Table 3 shows the average and maximum wind velocities measured for 26 days for the field measurement test models. The fence with a 4 m width (model B) showed a 65% decrease in the average wind velocity and a 49% decrease in the maximum wind velocity, when compared with external wind velocity. However, regarding the fence with a 6 m width (models A and B), the average and maximum wind velocity decreased by 47% and 23%, respectively, from the equivalents of external wind velocity.

Table 2: Specifications of the windbreaks installed in the weather station (unit = m).

	Width	Height	The distance from the windbreak wings to the ground
Model A	6	0.6	0
Model B	4	0.6	0
Model C	6	0.6	0.15

Table 3: Average and maximum wind velocities by the type of models (unit: m/s).

	Model A	Model B	Model C	Outside
Average	0.40	0.26	0.43	0.75
Maximum	1.49	0.99	1.48	1.93

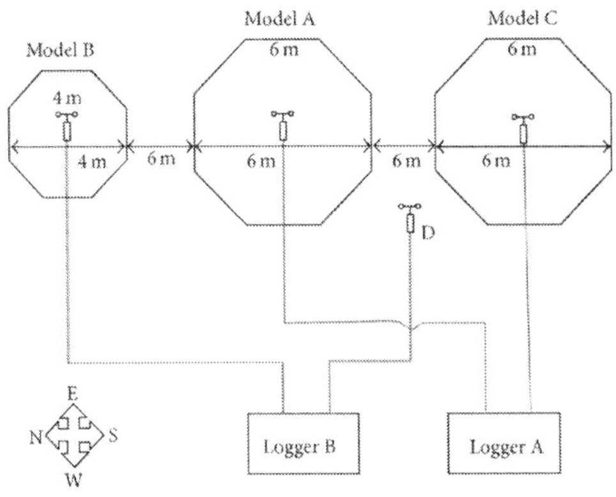

Figure 22: Arrangement plan and measurement process of anemometers.

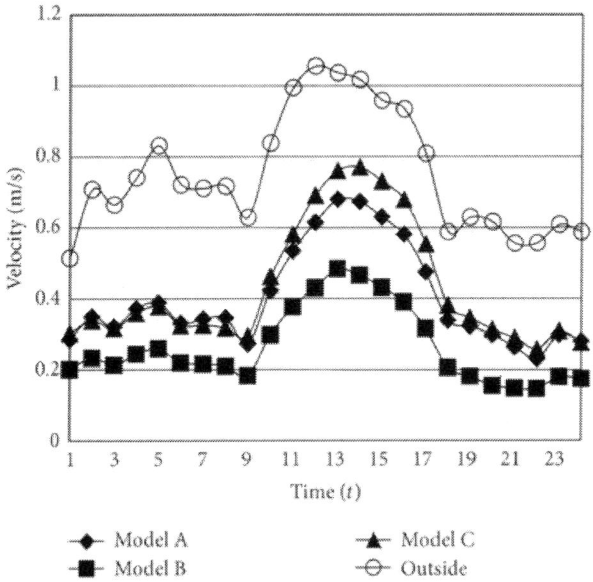

(a) Average wind velocity

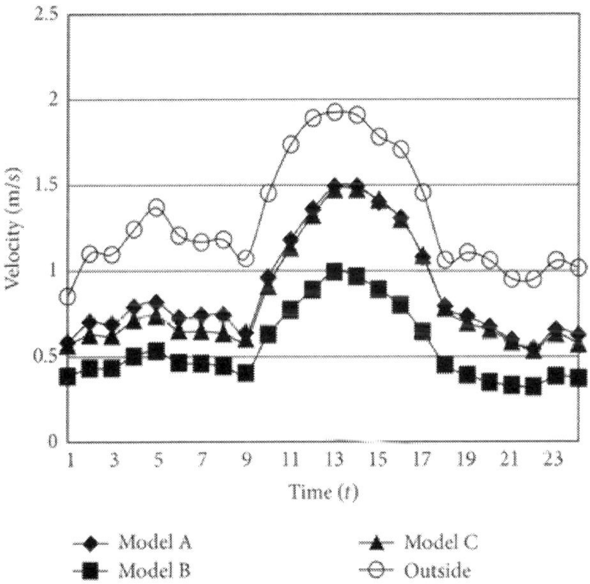

(b) 1 hour maximum wind velocity

Figure 23: Average and maximum wind velocities during each time.

CONCLUSION

Weather stations plates are set up on the flat ground in meteorological observatories, in order to measure the amount of snowfall. However, accurately measuring the amount of snowfall covering the ground is hard due to the influence of the wind. The results of the CFD analysis, wind tunnel test, and field measurement (aimed at examining changes in wind velocities within the fences, according to varying fence widths and heights) are as follows.

According to the result of the CFD analysis of the linear windbreak fences, a double fence whose vertical bars were fixed on the bottom is more effective in decreasing wind velocity. The distance where wind velocity decreased to a maximal extent varied according to the height of fences. The results of the CFD analysis, wind tunnel test, and field measurement of the octagonal windbreak fences showed that the wind velocity decrease rate was greatest at the center of the octagonal windbreak, when the fence width was 4 m and its height was 60 cm.

Conflict of Interests

The authors declare that there is no conflict of interests regarding the publication of this paper.

ACKNOWLEDGMENTS

This work was supported by the National Research Foundation of Korea (NRF) and was grant funded by the Korea government (MEST) (no. 2011-0028567).

REFERENCES

1. E. J. Plate, "The aerodynamics of shelter belts," Agricultural Meteorology, vol. 8, pp. 203–222, 1971.
2. M. D. A. E. S. Perera, "Shelter behind two-dimensional solid and porous fences," Journal of Wind Engineering and Industrial Aerodynamics, vol. 8, no. 1-2, pp. 93–104, 1981.
3. K. G. Ranga Raju, R. J. Garde, S. K. Singh, and N. Singh, "Experimental study on characteristics of flow past porous fences," Journal of Wind

Engineering and Industrial Aerodynamics, vol. 29, no. 1–3, pp. 155–163, 1988.

4. M. J. Judd, M. R. Raupach, and J. J. Finnigan, "A wind tunnel study of turbulent flow around single and multiple windbreaks, part I. Velocity fields," Boundary-Layer Meteorology, vol. 80, no. 1-2, pp. 127–165, 1996.
5. S.-J. Lee and H.-B. Kim, "Laboratory measurements of velocity and turbulence field behind porous fences," Journal of Wind Engineering and Industrial Aerodynamics, vol. 80, no. 3, pp. 311–326, 1999.
6. C. W. Park and S. J. Lee, "The effects of a bottom gap and non-uniform porosity in a wind fence on the surface pressure of a triangular prism located behind the fence," Journal of Wind Engineering and Industrial Aerodynamics, vol. 89, no. 13, pp. 1137–1154, 2001.
7. sc Stream Version 9. 0 Cradle co., Japan.
8. T.-H. Shih, W. W. Liou, A. Shabbir, Z. Yang, and J. Zhu, "A new k-ε eddy viscosity model for high Reynolds number turbulence flow," Journal of Wind Engineering and Industrial Aerodynamics, vol. 24, no. 3, pp. 227–238, 1995.
9. N. G. Wright and G. J. Easom, "Non-linear k-ε turbulence model results for flow over a building at full-scale," Applied Mathematical Modelling, vol. 27, no. 12, pp. 1013–1033, 2003.
10. B. E. Launder and D. B. Spalding, Lectures in Mathematical Models of Turblence, Academic Press, London, UK, 1972.
11. E. Simiu and R. H. Scanlan, Wind Effects on Structures, JohnWiley & Sons, 1996.
12. J. D. Holmes, Wind Loading on Structures, Spon, 2001.

Chapter 7

NUMERICAL ANALYSIS OF ELECTROMAGNETIC CONTROL OF THE BOUNDARY LAYER FLOW ON A SHIP HULL

Mohammad Bakhtiari, Hassan Ghassemi

Department of Ocean Engineering, Amirkabir University of Technology, Tehran, Iran

Email: navalarchitect2000@yahoo.com, gasemi@aut.ac.ir

ABSTRACT

In this article, electromagnetic control of turbulent boundary layer on a ship hull is numerically investigated. This study is conducted on the geometry of tanker model hull. For this purpose, a combination of electric and magnetic fields is applied to a region of boundary layer on stern so that produce wall parallel Lorentz forces in streamwise direction as body forces in stern flow. The governing equations including RANS equations with SST k-ω turbulent model coupled with electric potential equation are numerically solved by using Ansys Fluent codes. Accuracy of this turbulent model of Fluent in predicting Turbulent flow around a ship is also tested by comparing with available experimental results that it shows a good agreement with experimental data. The results obtained for ship flow show that by applying streamwise Lorentz forces that are large enough, flow is accelerated. The results are caused to delay or avoid the flow separation in stern, increase the propeller inlet velocity, create uniform flow distribution behind the ship's hull in order to improve the propeller performance, and finally decrease the pressure resistance and total resistance.

INTRODUCTION

The ability to manipulate a flow field to improve efficiency or performance is of immense technological importance. The intent of flow control may be to delay/advance transition, to suppress/ enhance turbulence, or to prevent/ promote separation. The resulting benefits include drag reduction, lift enhancement, mixing augmentation, heat transfer enhancement, and flow-induced noise. Flow control involves passive or active devices that have a beneficial change on the flow field. During the last decade, emphasis has been on the development of active control methods in which energy, or auxiliary power, is introduced into the flow. One of the active control methods is MHD control. If a fluid is electrically conducting, electromagnetic flow control permits to act directly within the boundary layer by applying directly local Lorentz forces.

Mainly three different force configurations have been investigated in order to control turbulent boundary layers: wall parallel streamwise, wall parallel spanwise and nominally wall normal forces. Wall parallel forces in streamwise direction have been applied, e.g., in the experiments of Henoch and Stace [1] and Weier et al. [2] as well as in the numerical analysis of Crawford and Karniadakis [3]. This force configuration increases instead of reducing wall shear stress, because the acceleration of the near wall fluid leads to a higher slope of the mean velocity profile in streamwise direction. However, the momentum gain due to the Lorentz force surpasses the friction drag rise. Nose chuck and Brown [4] used nominally wall normal, time dependent forces. Nose chuck and co-workers reported several successful experiments with a multitude of electromagnetic actuators generating turbulent skin friction reductions.

A circular cylinder equipped with electrodes as well as permanent magnets generating a wall parallel force in streamwise direction was used in the experiments and numerical calculations of Weier et al. [5]. Similar configurations have later been investigated by Kim and Lee [6], Posdziech und Grundmann [7], and Chen and Aubry [8]. While skin friction drag is increased by this force configuration, form drag is strongly reduced for an initially separated flow at Reynolds numbers Re = O(100). For stronger forcing, the increase in skin friction drag dominates the form drag decrease. Shatrov and Yakovlev [9] studied numerically the flow around a sphere with a mainly wall parallel Lorentz force at Re up to 1000. For increasing interaction parameter, i.e. the ratio of electromagnetic to inertial forces, the size of the separation region is first reduced. Later, separation is suppressed completely resulting in a strong decrease of form drag. Shatrov and Yakovlev [10] extended the investigated Reynolds number range up to 105 treating the problem of a steady and axially averaged flow. For large Reynolds numbers, the total drag of the sphere was reduced by 4 times and despite the moderate electrical efficiency of $\eta \approx 40\%$, the total energy consumption was reduced as well.

In present study, electromagnetic control of turbulent boundary layer on a ship hull, without free surface effects, is numerically investigated. For this purpose, a combination of electric and magnetic fields is applied to a region of Turbulent boundary layer on stern so that produce wall parallel Lorentz forces in stream wise direction as body forces in stern flow and their effects on flow parameters such as boundary layer thickness, flow separation, wake distribution on propeller plane, pressure distribution on ship hull and resistance is investigated. This study is done on the geometry of a tanker model hull. Very detailed experimental data are available for this ship [11]. For investigation of Lorentz forces effects, calculations are conducted for two cases without applying electromagnetic field and with applying electromagnetic field locally to stern flow. For this simulation, RANS equations with SST k-ω turbulent model coupled with electric potential equation are used. This governing equation is numerically solved in computational grid generated around ship hull by using Ansys Fluent 14.5 finite volume codes. Accuracy of SST k-ω turbulent model of Fluent in predicting Turbulent flow around a ship is also tested by comparing with available experimental results that it shows a good agreement with experimental data.

GOVERNING EQUATIONS

RANS Equations

In this study, it assumed that fluid is incompressible. The governing equations are the mass and momentum conservations. Using the Reynolds averaging approach, the Navies-Stokes equation can be stated as:

$$\frac{\partial u_i}{\partial x_i} = 0 \tag{1}$$

$$\frac{\partial}{\partial t}(\rho u_i) + \frac{\partial}{\partial x_j}(\rho u_i u_j) = -\frac{\partial p}{\partial x_i} + \frac{\partial}{\partial x_j}\left[\mu \frac{\partial u_i}{\partial x_j} - \rho \overline{u_i' u_j'}\right] + f_i \tag{2}$$

Where $-\rho\ u_i'u'j$ represents Reynolds stresses and f_i is Lorentz body force included in momentum equation for MHD flow

Turbulent Model

The Reynolds-averaged approach to turbulence modeling requires that the Reynolds stresses in Equation (2) are appropriately modeled. A common method employs the Boussinesq hypothesis to relate the Reynolds stresses to the mean velocity gradients:

$$-\rho \overline{u_i' u_j'} = \mu_t \left(\frac{\partial u_i}{\partial x_j} + \frac{\partial u_j}{\partial x_i}\right) - \frac{2}{3}\left(\rho k + \mu_t \frac{\partial u_k}{\partial x_k}\right)\delta_{ij} \tag{3}$$

In this article, the two-equation Shear-Stress Transport (SST) k-ω turbulence model is used for modeling turbulent viscosity.

The turbulence kinetic energy, k, and the specific dissipation rate, ω, are obtained from the following transport equations:

$$\frac{\partial}{\partial t}(\rho k)+\frac{\partial}{\partial x_i}(\rho k u_i)=\frac{\partial}{\partial x_j}\left[\left(\mu+\frac{\mu_t}{\sigma_k}\right)\frac{\partial k}{\partial x_j}\right]+\widetilde{G}_k-\rho\beta^* k\omega \tag{4}$$

$$\frac{\partial}{\partial t}(\rho\omega)+\frac{\partial}{\partial x_j}(\rho\omega u_j)=\frac{\partial}{\partial x_j}\left[\left(\mu+\frac{\mu_t}{\sigma_\omega}\right)\frac{\partial \omega}{\partial x_j}\right]+G_\omega-\rho\beta\omega^2+D_\omega \tag{5}$$

Where $\widetilde{G}_k$ represents the generation of turbulence kinetic energy due to mean velocity gradients:

$$\widetilde{G}_k=\min\left(-\rho\overline{u_i'u_j'}\frac{\partial u_j}{\partial x_i}, 10\rho\beta^* k\omega\right) \tag{6}$$

G_ω represents the generation of ω :

$$G_\omega=\frac{\alpha}{\nu_t}\widetilde{G}_k \tag{7}$$

σ_k and σ_ω are the turbulent Prandtl numbers for k and ω, respectively:

$$\sigma_k=\frac{1}{F_1/\sigma_{k,1}+(1-F_1)/\sigma_{k,2}} \tag{8}$$

$$\sigma_\omega=\frac{1}{F_1/\sigma_{\omega,1}+(1-F_1)/\sigma_{\omega,2}} \tag{9}$$

D_ω is cross-diffusion term. The turbulent viscosity, μ_t , is computed as follows:

$$\mu_t=\frac{\rho k}{\omega}\frac{1}{\max\left[\frac{1}{\alpha^*},\frac{SF_2}{\alpha_1\omega}\right]} \tag{10}$$

where S is the strain rate magnitude, the coefficient α^* damps the turbulent viscosity causing a low-Reynolds number correction .

1 F and F2 are blending functions.

Electric Potential Equation

In general, the electric field E can be expressed as:

$$E = -\nabla \phi - \frac{\partial A}{\partial t} \tag{11}$$

Where φ and A are the scalar potential and the vector potential, respectively. For a static field and assuming b << B_0 in which b is induced magnetic field and B_0 is imposed magnetic field, Ohm's law can be written as:

$$J = \sigma\left(-\nabla \phi + (U \times B_0)\right) \tag{12}$$

For sufficiently conducting media, the principle of conservation of electric charge gives:

$$\nabla \cdot J = 0 \tag{13}$$

The electric potential equation is thus given by:

$$\nabla^2 \phi = \nabla \cdot (U \times B_0) \tag{14}$$

The current density can then be calculated from Equation (12).

With the knowledge of the induced electric current, the MHD coupling is achieved by introducing additional source terms to the fluid momentum equation. This additional source term is the Lorentz force given by:

$$F = J \times B_0 \tag{15}$$

Which has units of (N/m^3) in the SI system.

For MHD flow around a ship hull, the magnetic field induced by the motion of sea water can be neglected compared with the imposed magnetic field. On the other hand, when the electric field is applied, a magnetic field is generated around the electric current. The magnitude of this generated magnetic field is evaluated by integrating Maxwell equation, $\nabla \times B = \mu j$, in the region where the electric field is applied:

$$B \sim \mu IS \tag{16}$$

Where I is total electric current and S is area of applied electric field. μ is of $O(10^{-5})$ for sea water, and assuming that I is of 103 Amperes at most, this induced magnetic field can again be neglected, in comparison with the imposed magnetic field. So potential electric equation in which the induced magnetic field is neglected, can be a proper method for MHD flow around a ship hull.

Finally, the governing equations to solve become seven equations including four RANS equations, two transport equations for k and ω and one potential equation that must be solved simultaneously.

COMPUTATIONAL GRID AND BOUNDARY CONDITIONS

In present study, a tanker model hull is used for calculate on of flow around ship. This model was presented at CFD workshop in Tokyo [11]. Very detailed experimental data of this ship is available for public use. Main dimensions of this model are shown in Table 1.Flow around the hull has been considered symmetrical to the center plane of the ship so the calculation is only conducted in on side of the symmetrical plane. Figure 1 shows the computational domain and boundary conditions applied for turbulent flow calculations. The distance of domain boundaries from the ship hull has been considered large enough in order to apply correct boundary conditions. These boundaries include the surface bellow:

1) A spherical surface with a radius of $R = 3L_{pp}$, that its center is on the intersection of aft perpendicular and water plane

2) A cylindrical surface with a radius R =3L_{pp}, that extends for 5L_{PP} from aft perpendicular to downstream flow.

For studying of electromagnetic effects on flow, an external magnetic field is applied to a region of flow limited to following boundaries (by assuming that the origin is on intersection of aft perpendicular and water plane):

- In longitudinal direction from x = 0.3 m plane to x = 1.3 m plane
- In transverse direction from hull surface to y = 0.5 m plane
- In vertical direction from z = −0.05 m plane to z = 0.25 m plane

Figure 2 represents MHD region and conditions applied to this region. These conditions are:

1) An external magnetic field of B_0 = 1T applied in z-axis negative direction.
2) "Conducting Wall" boundary condition with electric potential of φ0 = 20 V applied on ship hull in MHD region.

Since we have a large change in velocity in the wall normal direction in boundary layer, we need to use

Table 1. Principle dimensions of a tanker model.

Topic	Unit	Value
Length between perpendiculars	m	4.9700
Breadth (Molded)	m	0.9008
Draft (Molded)	m	0.3231
Wetted Surface Area	m3	6.5597
Displacement	m3	1.1712
Center of Buoyancy from Midship	%Lpp	3.50
Block Coefficient		0.8098

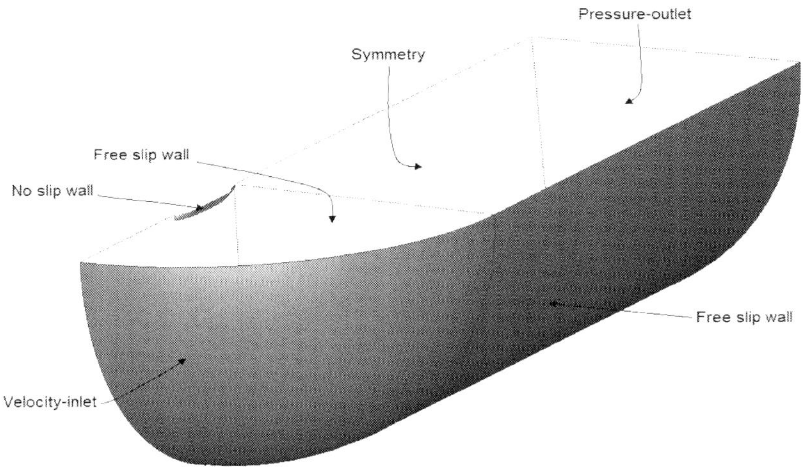

Figure 1. Solution domain and boundary conditions.

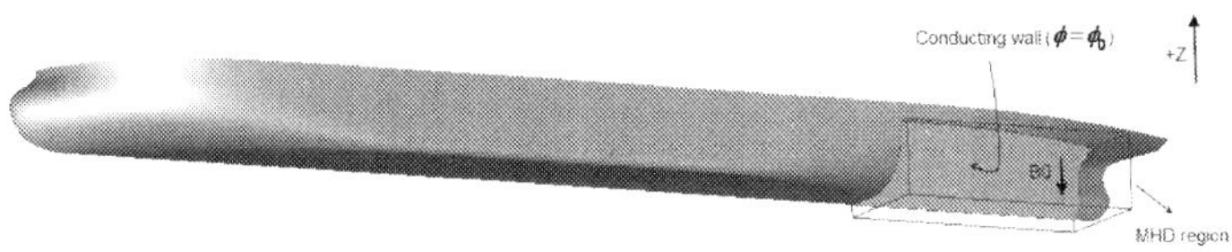

Figure 2. MHD region and conditions applied.

inflation layer meshing with high resolution in boundary layer region. In generating near wall mesh, distance of the first computational node from the wall and numbers of nodes placed within the boundary layer is very important and strongly depends on turbulent model used. We use y+ as non-dimensional distance of the first node from the wall.

For accurately predicting the velocity profile ,pressure gradients, boundary layer thickness and separation in turbulent boundary layer of ship and also resistance forces (that are of interest in present study), using a low-Reynolds turbulent model to solve whole boundary layer (including laminar sub-layer, Buffer region and fully turbulent layer) is needed. To ensure that we are capturing the laminar sub-layer, the average y+ value should be on the order of ~1 and the minimum number of nodes should be between 10 and 20 inside the boundary layer. So in this study, SST k-ω turbulent model is used and the whole solution domain is discretized by structured mesh including hexahedral cells. In order to ensure that the solution is independent of element size and the resolution of generated mesh is fine enough to obtain a numerical solution with minimum discretization error, calculations have been conducted in four computational mesh with different element sizes and cell numbers.

RESULTS AND DISCUSSIONS

As mentioned earlier, to investigate the effect of Lorentz forces, two cases were investigated. In the first case, no electromagnetic field was applied, while for the second case, an electromagnetic field was applied locally to stern flow. We utilized pressure based solver type with Spatial discretization of Body force weighted for pressure and spatial discretization of second order upwind for other field variables.

Calculations are conducted at Froude number Fn = 0.1425 and Reynolds number Rn = $4.8\times10^{6.}$ The working fluid of water has density ρ = 998.2 kg/m^3, viscosity μ = 0.001 kg/m·s and electrical conductivity σ = 5 s m .To ensure the convergence of solution, the iterations continues until the residual errors reduce to acceptable value of 10^{-4} and the total resistance coefficient, Cd, reaches a steady state and in this situation, the imbalance between inlet mass flow rate and outlet mass flow rate is less than 1%.

Results without MHD

In Table 2, the computed values of resistance coefficients, obtained without free surface effects, are shown. In same table, the total resistance coefficient measured experimentally is also shown. Note a good agreement between the results with a difference of 7%.

A comparison between the computational and experimental results for pressure coefficient distribution on stern is represented in Figures 3. It can be observed that there is a good agreement between computational and experimental results.

Furthermore, the results obtained from numerical computation for axial velocity distribution on the propeller plane compares with experimental results in Figure 4. Again, we can see that computed results show a good agreement with experimental ones

Results with MHD

In this section, the computed results with applying electric potential of φ_0 = 20 volts on ship hull in MHD region as a boundary condition, and an external magnetic field of B_0 = 1T in z-axis negative direction at this region (see Figure 2) are represented and then, the effects of Lorentz forces distribution caused by this applied conditions on turbulent boundary layer flow around ship hull is investigated.

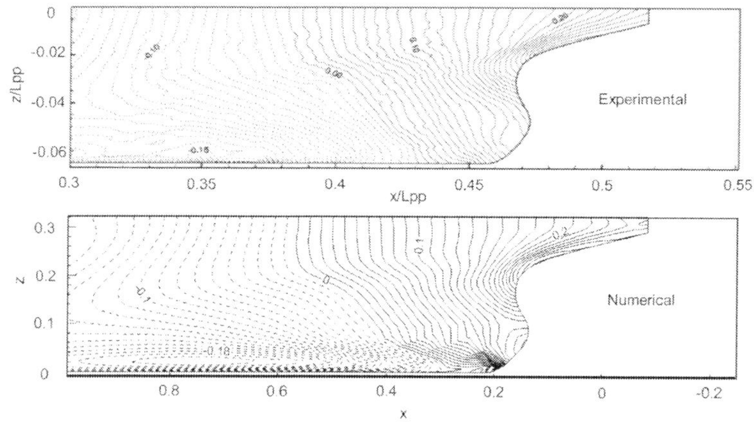

Figure 3. Comparison between experimental and numerical pressure coefficient distribution on stern (port side view).

Figure 5 shows the growth of boundary layer flow for two cases (with and without Lorentz forces). The comparison between these two cases demonstrates that the flow is accelerated due to Lorentz forces and boundary layer thickness decreases.

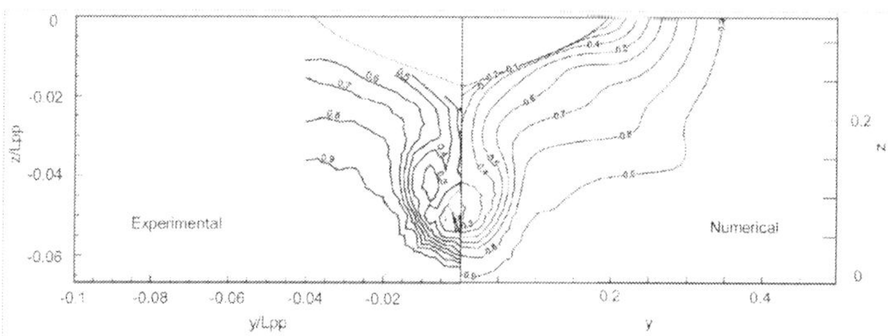

Figure 4. Comparison between experimental and numerical axial velocity distribution on the propeller plane.

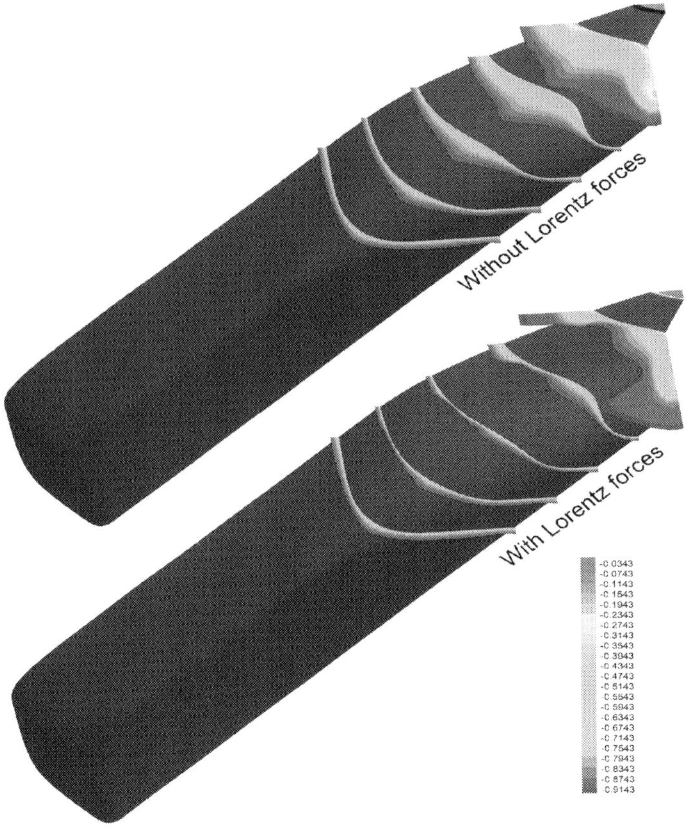

Figure 5. Comparison of boundary layer thickness for two cases (with and without Lorentz forces).

Furthermore, Figure 6 shows a comparison between two cases for axial velocity distributions computed on the propeller plane. From this comparison, it is clear that propeller inlet velocities have increased due to a decrease in boundary layer thickness and separation, which can lead to enhancement in propeller efficiency.

In Table 3, a comparison between computed values of resistances for two cases (with and without Lorentz forces) is shown.

CONCLUSIONS

In this article, a combination of electric and magnetic fields producing wall parallel Lorentz forces in stream wise direction as body forces, means which were applied to a region of turbulent boundary layer flow on stern of a ship model and its effects on flow around ship hull without free surface effects were numerically investigated.

With regard to the numerical results obtained from the calculations, the following conclusions can be offered:

- By applying wall parallel Lorentz forces in streamwise direction to a region of turbulent boundary layer flow on stern of a ship, we can decrease boundary layer thickness and flow separation on stern.
- This type of Lorentz forces can increase axial velocity on propeller plane and so the wake distribution on propeller is improved that leads to enhancement in propeller efficiency.
- The pressure on stern is significantly recovered due to presence of this Lorentz forces, so total ship resistance decreases.

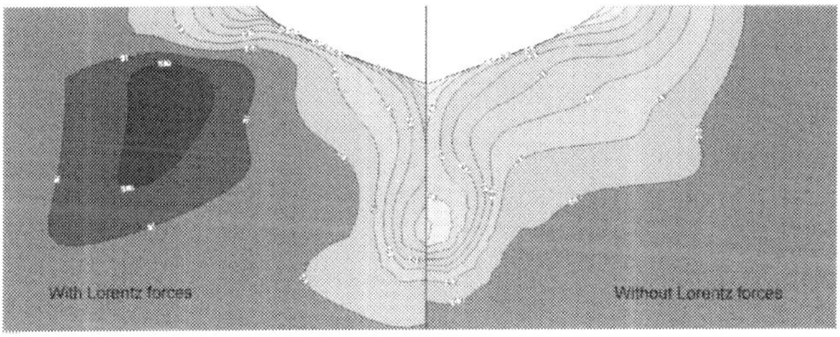

Figure 6. Comparison of axial velocity distributions on the propeller plane for two cases.

Table 2. Experimental and computational results for resistance coefficients.

Topic	Computational results	Experimental results	Relative Error
Pressure resistance coefficient	0.0028		
Frictional resistance coefficient	0.0136		

Total resistance coefficient	0.0164	0.0176	7%

Table 3. Computed resistances and their coefficients.

Topic	Without Lorentz forces	With Lorentz forces	Percent of decrease(−) or increase(+)
Pressure resistance (N)	2.1846	0.7686	−65%
Frictional resistance (N)	10.8166	11.4346	+5%
Total resistance (N)	13.0012	12.2032	−6%

- By using superconducting magnets, we can produce Lorentz forces that are large enough with a much Lower input MHD power. So by lowering the added power due to MHD system, the efficiency of this system can be increased.

REFERENCES

1. Henoch, C. and Stace, J. (1995) Experimental Investigation of a Salt Water Turbulent Boundary Layer Modified by an Applied Streamwise Magnetohydrodynamic Body Force. Physics of Fluids, 7, 1371-1383. http://dx.doi.org/10.1063/1.868525
2. Weier, T., Fey, U., Gerbeth, G., Mutschke, G., Lielausis, O. and Platacis, E. (2001) Boundary Layer Control by Means of Wall Parallel Lorentz Forces. Magnetohydrodynamics, 37, 177-186.
3. Crawford, C.H. and Karniadakis, G.E. (1997) Reynolds Stress Analysis of EMHD-Controlled Wall Turbulence: Part I Streamwise Forcing. Physics of Fluids, 9, 788-806. http://dx.doi.org/10.1063/1.869210
4. Nosenchuck, D. and Brown, G. (1993) Discrete Spatial Control of Wall Shear Stress in a Turbulent Boundary Layer. In: So, R., Speziale, C. and Launder, B., Eds., Near-Wall Turbulent Flows, Elsevier, Amsterdam.
5. Weier, T., Gerbeth, G., Mutschke, G., Platacis, E. and Lielausis, O. (1998) Experiments on Cylinder Wake Stabilization in an Electrolyte Solution by Means of Electromagnetic Forces Localized on the Cylinder Surface. Experimental Thermal and Fluid Science, 16, 84-91. http://dx.doi.org/10.1016/S0894-1777(97)10008-5
6. Kim, S. and Lee, C. (2000) Investigation of the Flow around a Circular Cylinder under the Influence of an Electromagnetic Force. Experiments

in Fluids, 28, 252-260. http://dx.doi.org/10.1007/s003480050385

7. Posdziech, O. and Grundmann R. (2001) Electromagnetic Control of Seawater Flow around Circular Cylinders. European Journal of Mechanics, 20, 255-274. http://dx.doi.org/10.1016/S0997-7546(00)01111-0
8. Chen, Z. and Aubry, N. (2005) Active Control of Cylinder Wake. Communications in Nonlinear Science and Numerical Simulation, 10, 205-216. http://dx.doi.org/10.1016/S1007-5704(03)00128-X
9. Shatrov, V. and Yakovlev, V. (1985) Hydrodynamic Drag of a Ball Containing Aconduction-Type Source of Electromagnetic Fields. Journal of Applied Mechanics and Technical Physics, 26, 19-24. http://dx.doi.org/10.1007/BF00919618
10. Shatrov, V. and Yakovlev, V. (1990) The Possibility of Reducing Hydrodynamic Resistance through Magnetohydrodynamic Streaming of a Sphere. Magnetohydrodynamics, 26, 114-119. [11] Hino, T. (2005) CFD Workshop Tokyo 2005. National Maritime Research Institute, Tokyo.

Chapter 8

3D THERMO-FLUID DYNAMIC SIMULATIONS OF HIGH-SPEED-EXTRUDED STARCH BASED PRODUCTS

Alessandro Cubeddu[1,2], Cornelia Rauh[1,2], Antonio Delgado[1]

[1]Institute of Fluid Mechanics, Technical Faculty, University of Erlangen-Nuremberg, Erlangen, Germany

[2]Department of Food Biotechnology and Food Process Engineering, Faculty III—Process Science and Engineering, Berlin Institute of Technology, Berlin, Germany

ABSTRACT

This paper aims to investigate a method to perform non-isothermal flow simulations in a complex geometry for generalized Newtonian fluids. For this purpose, 3D numerical simulations of starch based products are performed. The geometry of a co-rotating twin-screw extruder is considered. Process conditions concern high rotational speed (up to 1800 rpm), different flow rates (30, 40 and 60 kg/h) and water contents (22% and 36%), for a total of 54 simulations. To cope with the geometry complexity a Mesh Superposition Technique (MST) was adopted. The pseudo plastic behavior of the fluid is taken into account by considering viscosity as function of shear rate (Ostwald Wale relationship) and temperature (Arrhenius law). Simulated temperature variations are compared with measurements at same process conditions for validation. Qualitative behaviour of temperature T and shear stress $\tau\tau$ along the screw are analyzed and comparisons of different process conditions are presented. By these simulations a database is formed to develop a process control strategy for novel extruder operating points in food technology.

INTRODUCTION

Since their massive development in the first half of 20th century, extruders have been experiencing a non-stop growth in all industrial fields, not only in plastic industry but also in food processing. The need to develop new products and increase the existent production encourages manufacturers to develop new geometrical screw configurations and to promote research to better understand the complex nature of extrusion processes. Especially in the area of breakfast cereals and convenience food many works have been carried out in order to explain the rheological properties of starch based extrudates. Senouci and Smith [1] as well as Xie et al. [2] characterized shear stress and melt viscosity as function of product temperature, moisture content and melt amylose/ amylopectin content and investigated the regression power law parameters. The influence of the specific mechanical energy (SME) on cornmeal viscosity was illustrated by Chang et al. [3] while product degradation and expansion mechanisms were widely studied in many other works [4] [5] [6] [7]. An exhaustive review of the state of the art about the rheology of starch polymers and related rheometric techniques can be found in [8]. Engineering aspects of extruder performances from an energy point of view have been also investigated [9] [10] and assessment of twin-screw extrusion in producing snack food [11] as well as effects on product characteristics [12] have been experimentally studied. Some of the most important independent variables in the extrusion process are: preconditioning, screw speed N, water content xw, flow rate Q& , barrel temperature TW, die geometry and extruder configuration [13]. All manipulable variables influence in a non-linear way the mechanical and thermal stresses and are responsible for the quality of the final product. Due to the complexity of the extrusion process and high number of design variables, most research has been historically carried out on an experimental basis and only recently, with the growth in computer science, the empirical approach has been supported by numerical contributions. Since the begin of the 1990s theoretical models have been used to simulate the transformation of starchy twin-screw extrudates [14], but only with the advent of the new millennium first attempts have been done to solve the governing equations of fluid dynamics in 3D twin-screw geometries by means of Finite Element Method (FEM). Ishikawa et al. [15] modelled six discrete angular positions of the screw geometry and performed a non-isothermal quasi-steady-state simulation of a polypropylene material. Avalosse et al. [16] reproduced the results of Ishikawa using a Mesh Superposition Technique (MST). The MST as well as the Mesh Partitioning Technique [17] allows simulating the extrusion process without regenerating the finite element mesh as the screws rotate. Use of MST to perform isothermal fluid dynamic simulations of extruded starch based matrix can be found in [18].

The present work is intended to provide a method to set non-isothermal flow simulations in a complex geometry for generalised Newtonian fluids. The new contribution is the analysis and discussion of the flow behavior of the studied extrude comparing a wide range of process conditions, which cover new and extremely high operating points, up to 1800 rpm. For this aim the MST implemented in the POLYFLOW® software is applied to a twin-screw co-rotating geometry and non-isothermal 3D numerical simulations of starch based products are preformed. The analysis of the extrudate flow behaviour is carried out from a thermal and kinematic point of view for screw speeds ranging from 400 rpm to 1800 rpm. Three different feed rates (30, 40 and 60 kg/h) and two water contents (22% and 36%) are also considered for a total of 54 simulations. The modelled zone is the melt-conveying zone of the extruder, before the die, where the mixture flows as molten monophasic dough. Counter pressure is taken into account. Material parameters were provided by the Department of Food Process Engineering and Department of Applied Mechanics of Karlsruhe Institute of Technology, that performed the experiments and measurements. Validation is made with experimental data for the temperature across the simulated area. Profiles of temperature and shear stress along the simulated area are qualitatively analyzed. Quantitative comparisons of all process conditions are also discussed. The performed simulations investigate novel extrusion process conditions in food technology and offer a base of numerical data which can be used to develop a control strategy for optimisation and control of the extrusion process.

MATERIALS AND METHODS

Geometry and Simulated Area

In Figure 1 the extruder cross section is schematically represented. The modelled area is found at the end of the extruder (melt-conveying zone). The advantages of simulating the flow just before the die are a filling degree of 100%, the possibility to take into account the counter pressure and to compare the calculated temperature variations with the measurements. In fact the temperature transducer plugs are situated across the considered elements. The simulated area consists of three screw elements in a row. Frontal and lateral view with dimensions (inmm) of a single screw and the lateral view of the whole simulated domain are illustrated in Figure 2.

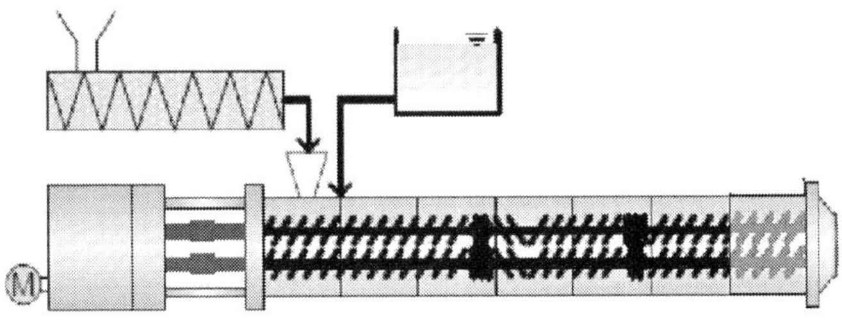

Figure-1 Schematic draw of the extruder and location of the modeled area.

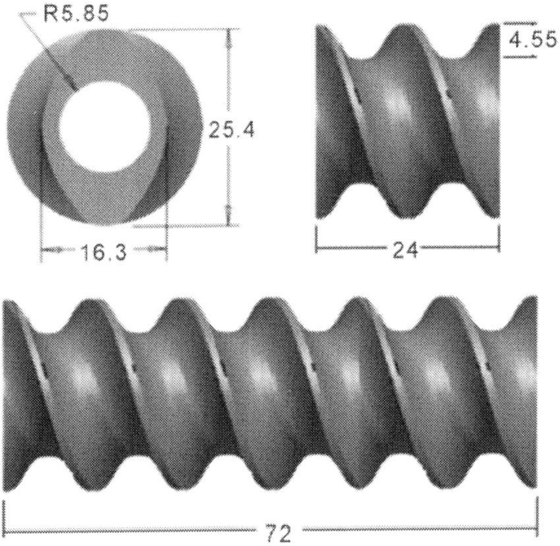

Figure-2. Frontal and lateral view of a single screw element and lateral view of the modeled area (3 elements). Dimensions are in mm.

Mesh Superposition Technique

Even in a steady state operating condition (i.e. constant screw speed and far away from start-up or shut-down transitories), due to the rotation of the helicoidal screw geometry, the fluid inside the extruder flows at a nonsteady periodic condition. The state variables are time dependent with a periodicity depending on the screw speed. Thus, the discretisation of the fluid domain is in principle also time dependent. In a classical approach [19] the mesh is created only on the volume occupied by the fluid, following the geometrical shape of the screw surface at different angles. As the screws rotate, the mesh must be regenerated to take into account the new relative position between the screw volume and the fluid domain. In this study the use of MST allows avoiding a re-meshing algorithm. The screw meshes and the fluid mesh are independently generated with the mesh generator program ICEM CFD®. The fluid mesh is made up of hexahedra bricks for a total of 25,230 elements. The screws are discretised with tetrahedral, for a total of 21,846 elements. The created meshes are then merged using the tool POLYFUSE, which allows to combine multiple meshes. In the resulting merged mesh some portions of the fluid overlap the screw cells (Figure 3). As the screw rotate new portions of the fluid domain will overlap partially or totally the screws. Making use of the MST requires a modification of the fundamental equations of fluid dynamics.

For an incompressible fluid, mass (Equation (1)), momentum (Equation (2)) and energy (Equation (3)) conservation equations can be written as follows[1] [20]:

$$\nabla \cdot v + \frac{\beta}{\eta}\nabla^2 P = 0 \quad (1)$$

where v is the velocity β is the relative compression factor, artificially added to guarantee the mass

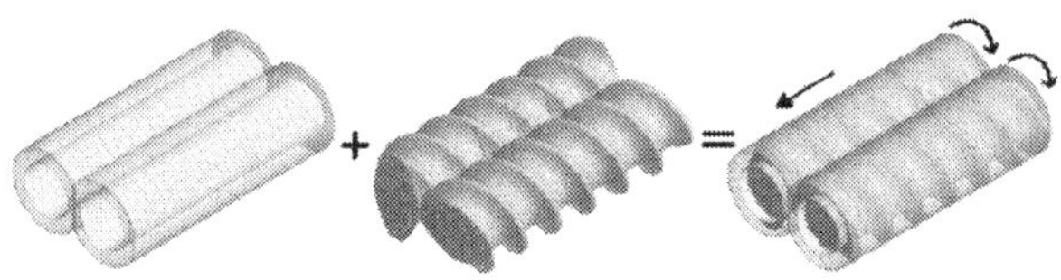

Figure-3 Domain discretisation and mesh superposition.

conservation and avoid peaks of pressure near to the moving elements, η is the shear viscosity and P the pressure.

$$H(v - v_{el}) + (1 - H)\left(\rho\frac{\partial v}{\partial t} + \rho v \cdot (\nabla v) + \nabla P - \nabla \cdot \bar{\bar{\tau}} + \rho g\right) = 0 \quad (2)$$

where $H \in \{0,1\}$ is a step function introduced to distinguish if a fluid mesh node is inside or outside the screw subdomain. If $H = 0$ then the node is considered part of the fluid, else it is considered to be inside of the moving parts and takes all physical properties of the screws.

The parameter ρ is the fluid density $\bar{\bar{\tau}}$ is shear stress tensor and g is the acceleration of volume forces,

$$H\left[\rho_{d}c_{p_{d}}\frac{DT}{Dt}-\nabla\left(k_{d}\nabla T\right)\right]+\left(1-H\right)\left[\rho c_{p}\frac{DT}{Dt}-\bar{\bar{\tau}}:\nabla v-\nabla\cdot\left(k\nabla T\right)\right]=0 \qquad (3)$$

where p_c is the specific heat capacity, T is the temperature and k is the thermal conductivity. D/Dt is the substantial derivative.

Kinematic and Thermal Boundary Conditions

The kinematic and thermal boundary conditions are shown in Figure 4. Rotation of the screws and main flow direction are indicated by the arrows. Flow rate Q& is imposed at the inlet domain. Simulations at 30, 40 and 60 kg/h have been performed. The chosen values of flow rate guarantee a filling degree of 100% in the considered area. Temperature at the inlet is in principle different for each process condition. Indeed, the inlet temperature decreases with the increase in water content and screw speed. However, for simplicity, we used two constant values of the inlet temperature: 418 K for all simulations at x_w =22% = and 373 K for all simulations at x_w =36%These values are experimental averages for each of the considered water contents, respectively.

On the outlet of the domain a counter pressure is imposed. The values of the counter pressure have been measured at the Karlsruhe Institute of Technology for each considered operating point. The screws are insulated solids which rotate at constant angular velocity. Screw speed varies from 400 rpm to 1800 rpm. A non-slip condition is imposed on the screw surface. On the barrel, there is a non-slip condition and a constant temperature of 373 K. The Newton´s law of cooling has been used on the boundary layer of the internal barrel surface in order to take into account convective effects:

$$-k\frac{\partial T}{\partial r}=h\left(T-T_{W}\right) \qquad (4)$$

where r is the radial direction and h is the heat transfer coefficient. Common value of h for starch based products is 500 W/(m^2·K) [21].

Material Parameters

The fluid is a non-Newtonian shear thinning fluid. It means that the higher the shear rate is, the smaller is the viscosity. A pure viscous model can be written as:

$$\eta = f(\dot{\gamma}, T) \tag{5}$$

where γ is the shear rate. The model used in this study is the Ostwald-de Waele relationship:

$$\eta(\dot{\gamma}, T) = K(T)\dot{\gamma}^{n-1} \tag{6}$$

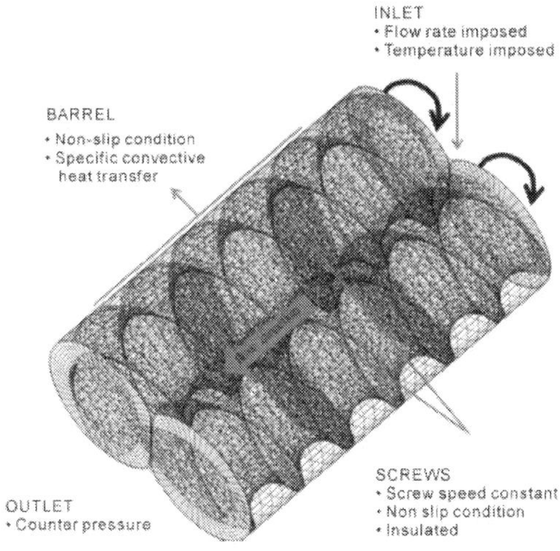

Figure 4 Kinematic and thermal boundary conditions.

where K (T) is the Arrhenius law and n is the flow index, representing the slope of the viscosity curve in a ln-ln diagram. It takes a value <1 in case of a shear thinning fluid like the one considered in this study. The Arrhenius law can be written as follows:

$$K(T) = Ae^{\alpha\left(\frac{1}{T}-\frac{1}{T_W}\right)} \quad (7)$$

where A is the consistency related to the magnitude of the average viscosity. It represents the apparent viscosity at unit shear rate. The parameter α is the ratio of the activation energy to the universal gas constant. In this case α takes the value of 2407 K, as the measured average activation energy is 20 kJ/mol. The complete equation for the viscosity can be written as follows:

$$\eta(\dot{\gamma},T) = Ae^{\alpha\left(\frac{1}{T}-\frac{1}{T_W}\right)}\dot{\gamma}^{n-1} \quad (8)$$

The flow function η (γ,T) is univocal for each process condition. The values of the consistency A and flow index n depend strongly on the process conditions. In general A decreases when screw speed N or water content w_x are increased, whereas n has the opposite behaviour.

The fluid density ρ is constant during the extrusion process and takes the value of 1351 kg/m^3 for w_x= 22%and 1288 kg/m3 for w_x = 36%.

Inertial terms and acceleration of gravity have been neglected: the inertial terms because the Reynolds number is less than one, due to the high fluid viscosity and small dimensions of the extruder section, and the acceleration of gravity because it does not play a relevant role in the extrusion process which takes place in a horizontal plane. The thermal conductivity of the fluid k is considered constant with a value of 0.6 W/(m·K) [22]. The fluid specific heat capacity p c has a constant value of 2300 J/(kg·K) [23].

Numerical Solution

Due to the high number of finite volumes and the complexity of the numerical problem, the matrices of the coefficients are very large sparse matrices whose factorisation requires high computational resources. For that reason a domain parallelisation was adopted, in order to reduce the computational load on the CPU. Therefore a mesh decomposition was made by using the Chaco decomposition method. A number of 512 sub-parts was specified to partition the whole domain in smaller and simpler to handle sub-domains. The resulting sub-matrices are then solved and reassembled on a parallel basis making use of a High Performance Computing (HPC) technology. To take advantage of the HPC parallelisation capabilities, a multifrontal method [24] has been adopted as solver. To avoid convergence problems due to the dependence of the viscosity on both shear rate and temperature, two simulations for each process condition were performed. In the first simulation only the viscosity dependence on shear rate is considered, while $K(T) = 1$. Momentum and energy equations are hence decoupled. In the second simulation the results of the first simulation are used as initial conditions and only the viscosity dependence on temperature is taken into account. In this case momentum and energy are coupled and therefore an iterative one-way coupled solution is adopted. The one-way method ensures convergence of the solution even if at a convergence rate generally lower than a full coupled method. Convergence difficulties can be caused also by the high value of Péclet number. It is defined as the ratio of the advection heat transfer to the conduction heat transfer:

$$Pe = \frac{\rho c_p d v}{k} \tag{9}$$

where d is a characteristic dimension of the geometry (in this case the depth of the channel, d = 4.55 mm) and v is the modulus of the flow velocity. In polymer extrusion the advection heat transfer term is much higher than the conduction heat transfer. The Péclet number is of the order of 103 to 105 [25]. In this study it varies from 3×10^3 to 3×10^4, depending on the process conditions. A high Péclet number means high non-linearity of the energy equation and causes instability of the numerical solution. To solve this problem the Péclet number was varied iteratively during the second step simulation from a very small value (<<1) to its actual value. Another dimensionless number common in extrusion science is the Brinkman number. It is defined as the ratio of viscous heat dissipation to heat conduction transfer through the barrel:

$$Br = \frac{\eta v^2}{k\left|T_W - T_{ref}\right|} \tag{10}$$

where T_{ref} is a reference temperature, e.g. the bulk temperature of the fluid. High values of Brinkman number are very common in the extrusion process, due to the high viscosity dissipation. This is a cause of numerical instability. In this study, the Brinkman number ranges from 10 to 30 and therefore also for this number an iterative evolution (from a very small value to its actual value) was adopted.

RESULTS AND DISCUSSION

In Table 1 are summarised the Ostwald-de Waele parameters A and n of 8 numerical simulations used to validate the temperature field. A constant flow rate of 40 kg/h has been used. Of the 8 validation points, the first 4 are referred to a fluid with 22% of water content and the remaining 4 points to a fluid with 36% of water content. The comparison between measured ΔT and simulated ΔT across the modelled domain are diagrammed in Figure 5. Simulation results show a good agreement with the measured values for both considered water contents. The source of error is due to the simplifications adopted in the simulations. For instance, the inlet temperature is considered function of the only water content because this variable influences considerably the temperature at the inlet. The influence of the screw speed on the inlet temperature is neglected. Furthermore, thermal parameters like specific heat capacity, thermal conductivity and convective heat transfer coefficient are kept constant, for simplicity. However, the simulations can reproduce accurately the temperature variation trend, demonstrating the suitability of the Ostwald-de Waele model to reproduce the extrudate temperature.
State Variables along the Screw

It is interesting to focus the attention on one simulation (N = 800 rpm, x_w 36% = , Q= 40 kg/h) and have a look at the qualitative variation of the temperature and shear stress in the simulated area. In Figure 6 it is represented the temperature contour over two revolution surfaces located at 11 mm of radius from the rotational axes and over the plane of symmetry of the extruder. The screws rotate anticlockwise and the fluid is pumped in z-direction. The temperature contour shows an increasing in temperature in z-direction from the

Table 1 Ostwald-de Waele parameters A and n used to validate the model.

Q (kg/h)	x_w (%)	N (rpm)	A (Pa·s^n)	n
40	22	600	5757	0.404
40	22	1000	3058	0.465
40	22	1400	1635	0.556
40	22	1800	809	0.656
40	36	600	1570	0.398
40	36	1000	850	0.532
40	36	1400	436	0.648
40	36	1800	197	0.763

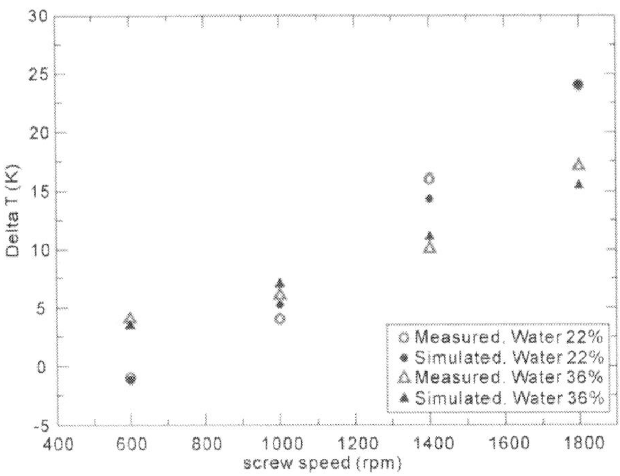

Figure 5 Comparison between simulated and measured values of temperature.

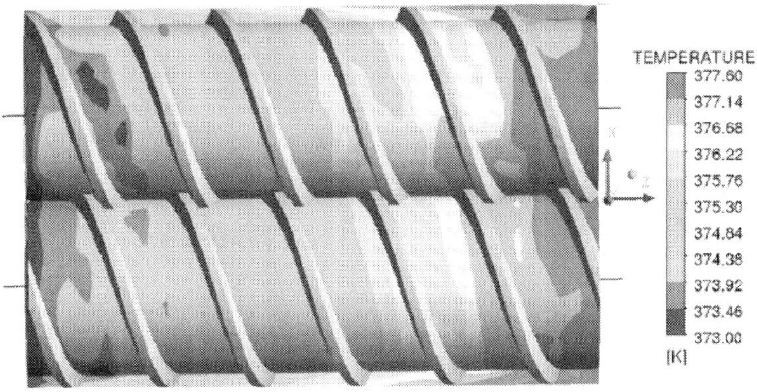

Figure 6 Temperature contour in the simulated area.

initial value of 373 K to its final value at the outlet of the domain. The viscous dissipation is the cause of the temperature increase. As the temperature increases, the viscosity decreases (shear thinning behaviour) resulting in a reduction of viscous dissipation and then in a lower increase of the temperature. The equilibrium is reached when the viscous heating is balanced by the heat flux through the barrel. The dissipative effects are homogeneously distributed and the temperature profile shows an almost linear increasing trend along the screw (Figure 7). The shear stress contour is represented in Figure 8. Highest values of shear stress are found in the intermeshing zone (between the two screws) and especially close to the barrel. Here the stresses.

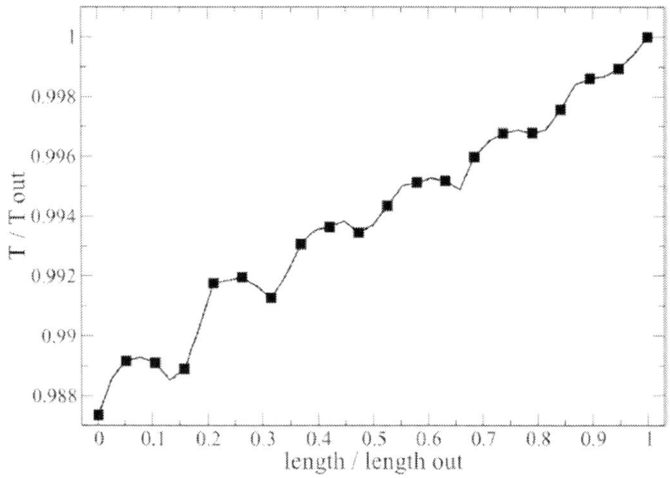

Figure 7 Normalised temperature variation along the screw.

Figure 8 Shear stress contour in the simulated area.

concentrate because of the high relative speed between screws (intermeshing zone) and between screws and barrel (close to the barrel wall). In z-direction the shear stress presents a periodic behavior. This can be seen in Figure 9 that shows the normalised shear stress variation in a line parallel to the rotation axes. In the middle of each channel where the velocity gradient is higher, the shear stress has a higher value in comparison with its value near to the flights.

Comparison of Different Process Conditions

The fluid dynamic variables ΔT and average shear stress in the simulated area in function of screw speed are diagrammed in figures 10 to 13. Figure 10 and Figure 11 show a comparison of the temperature variations for the three flow rates considered in this study, i.e. 30, 40 and 60 kg/h. The water content of the dough is 22% and 36% for Figure 10 and Figure 11, respectively. For all curves there is an increase of ΔT with the screw speed. This is due to the higher mechanical energy transferred to the fluid at higher screw speed. The curves show in general a reduction of the ΔT while the flow rate is increasing. In fact, the higher the flow rate is, the lower is the permanence time of the dough inside the pumping area and then the temperature increase. Comparing Figure 10 with Figure 11, it is to be noted that, fixing the other process conditions, the temperature increase is smaller for dough with higher percentage of water content in comparison with a fluid with a lower water content. This is the consequence of the smaller average viscosity of a fluid with higher water content and therefore a minor viscous dissipation. The curve at 60 kg/h shows a smaller slope in respect to the curves at lower flow rates. It may be caused from the fact that a high reduction of the permanence time of the dough inside the pumping area makes the fluid less sensitive to the other process conditions.

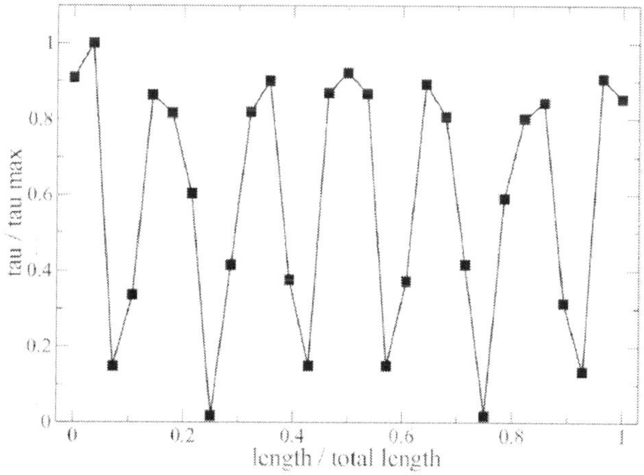

Figure 9 Normalized shear stress along the screw.

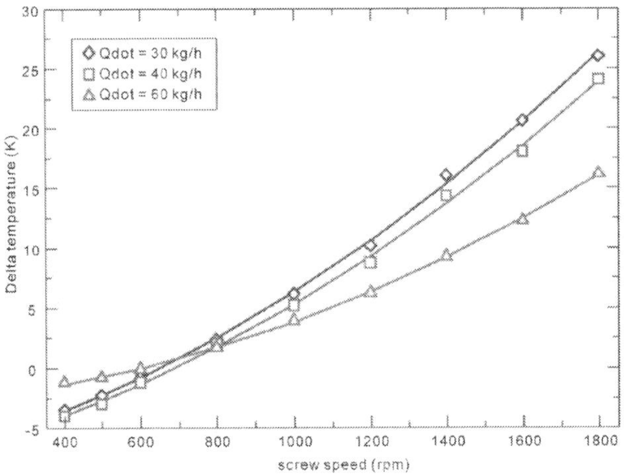

Figure 10 ΔT in function of screw speed for three different flow rates. Water content is 22%.

Therefore an increase in screw speed results in a smaller percentage of temperature increase, which causes a flattening of the curve.

In Figure 12 and Figure 13 the average shear stresses at different process conditions are compared for an extrudate with 22% and 36% of water content respectively. In case of a dough with medium water content (Figure 12), the shear stress decreases while the screw speed is increasing, whereas in case of a dough with high water content (Figure 13) the shear stress increases while the screw speed is increasing. This can be explained with a different behaviour of the viscous fluid at different water contents. Increasing the water content results in fact in a more Newtonian behaviour of the fluid. Therefore, the effects of mechanical stress on the fluid, caused by the increasing in screw speed, overtakes the effects of the viscosity reduction due to the shear thinning nature of the fluid. Thus the shear stress increases with the screw speed. At medium and

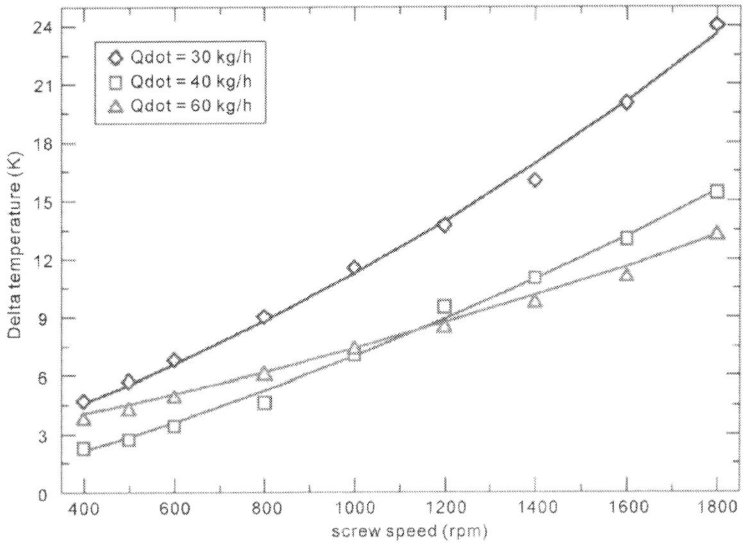

Figure 11 ΔT in function of screw speed for three different flow rates. Water content is 36%.

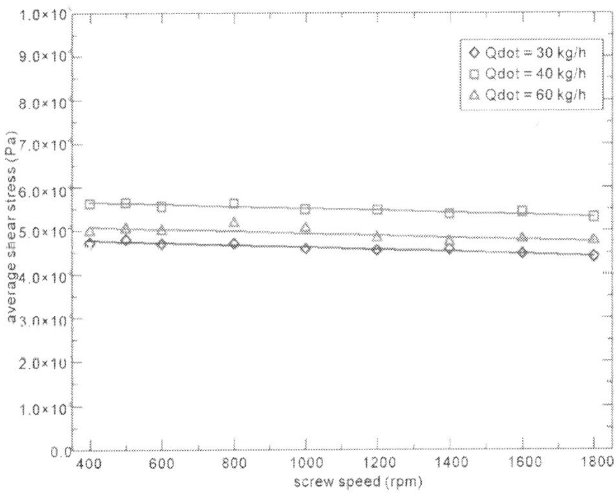

Figure 12 Average shear stress in function of screw speed for three different flow rates. Water content is 22%.

low water contents the shear thinning behaviour of the polymer is vice versa more relevant and the shear stress reduces while the screw speed is increasing. The influence of the flow rate on the shear stress seems not to be very relevant. The absolute values of shear stresses remain of the order of 50 kPa and 20 kPa for a fluid with 22% and 36% of water content respectively. The higher absolute values for the extrudate with at 22% of water content is due to its higher average viscosity in comparison with a fluid at 36% of water content.

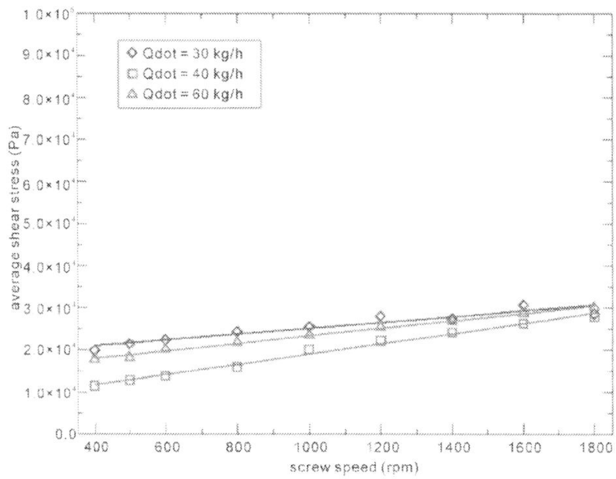

Figure 13 Average shear stress in function of screw speed for three different flow rates.

Water content is 36%.

CONCLUSION

3D thermo fluids dynamic simulations of starch based extruded products have been performed. 54 numerical simulations have been carried out for a wide range of process conditions considering novel food properties, different flow rates and very high screw speeds. The fluid has been modelled with the Ostwald-de Waele relationship and the Arrhenius law. The study shows that the chosen viscous model coupled with the temperature dependence of the viscosity is able to describe the behaviour of the studied dough inside the pumping area of an industrial extruder. Temperature variation as well as shear stress is discussed for a fixed operating point in the simulated area and along the pumping direction. Comparison of the simulated variables is also presented for all studied process conditions and interpretations are proposed. Results of the simulations can be used in the industry and for design of a process control strategy which allows to optimise the process of extrusion and to improve the product quality.

ACKNOWLEDGEMENTS

This research was supported by the German Ministry of Economics and Technology (via AiF) and the FEI (Forschungskreis der Ernährungsindustrie e.V., Bonn) Project AiF 332 ZN. Thanks to the company Coperion GmbH for providing the screw geometry.

REFERENCES

1. Senouci, A. and Smith, A.C. (1988) An Experimental Study of Food Melt Rheology. I. Shear Viscosity Using a Slit Die Viscometer and Capillary Rheometer. Acta Rheologica, 27, 546-554.
2. Xie, F., Yu, L., Su, B., Liu, P., Wang, J. and Chen, L. (2009) Rheological Properties of Starches with Different Amylose/ Amylopectin Ratios. Journal of Cereal Science, 49, 371-377.
3. Chang, Y.K., Martinez-Bustos, F., Park, T.S. and Kokini, J.L. (1999) The Influence of Specific Mechanical Energy on Cornmeal Viscosity Measured by an On-Line System during Twin-Screw Extrusion. Brazilian Journal of Chemical Engineering, 3, 285-295.
4. Willet, J.L., Millard, M.M. and Jasberg, B.K. (1997) Extrusion of Waxy Maize Starch: Melt Rheology and Molecular Weight Degradation of Amylopectin. Polymer, 38, 5983-5989.
5. Della Valle, G., Boché, Y., Colonna, P. and Vergnes, B. (1995) The Extrusion Behaviour of Potato Starch. Carbohydrate Polymers, 28, 255-264.
6. Della Valle, G., Vergnes, B., Colonna, P. and Patria, A. (1997) Relations between Rheological Properties of Molten Starches and their Expansion Behaviour in Extrusion. Journal of Food Engineering, 31, 277-296.
7. Launay, B. and Lisch, J.M. (1983) Twin-Screw Extrusion Cooking of Starches: Flow Behaviour of Starch Pastes, Expansion and Mechanical Properties of Extrudates. Journal of Food Engineering, 2, 259-280.
8. Xie, F., Halley, P.J. and Avérous, L. (2012) Rheology to Understand

and Optimize Processibility, Structures and Properties of Starch Polymeric Materials. Progress in Polymer Science, 37, 595-623.

9. Janssen, L.P.B.M., Mościcki, L. and Mitrus, M. (2002) Energy Aspects in Food Extrusion-Coocking. International Agrophysics, 16, 191-195.
10. Liang, M., Huff, H.E. and Hsieh, F.H. (2001) Evaluating Energy Consumption and Efficiency of a Twin-Screw Extruder.Journal of Food Science, 67, 1803-1807.
11. Shi, C., Wang, L., Wu, M., Adhikari, B. and Li, L. (2011) Optimization of Twin-Screw Extrusion Process to Produce Okara-Maize Snack Foods Using Response Surface Methodology. Journal of Food Engineering, 7, 1-24.
12. Guha, M., Ali, S.Z. and Bhattacharya, S. (1997) Twin-Screw Extrusion of Rice Flour without a Die: Effect of Barrel Temperature and Screw Speed on Extrusion and Extrudate Characteristics. Journal of Food Engineering, 32, 251-267.
13. Hubner, G.R. (2000) Twin-Screw Extruders. In Riaz, M.N., Ed., Extruders in Food Applications, CRC Press, Taylor & Francis Group, Boca Raton, 8-114.
14. Della Valle, G., Barrès, C., Plewa, J., Tayeb, J. and Vergnes, B. (1993) Computer Simulation of Starchy Products' Transformation by Twin-Screw Extrusion. Journal of Food Engineering, 19, 1-31.
15. Ishikawa, T., Kihara, S.I. and Funatsu, K. (2000) 3-D Numerical Simulations of Nonisothermal Flow in Co-Rotating Twin Screw Extruders. Polymer Engineering and Science, 40, 357-364. http://dx.doi.org/10.1002/pen.11169.
16. Avalosse, T., Rubin, Y. and Fondin, L. (2002) Non-Isothermal Modeling of Co-Rotating and Contra-Rotating Twin Screw Extruders. Journal of Reinforced Plastics and Composites, 21, 419-429.
17. Gupta, M. (2008) Non-Isothermal Simulation of the Flow in Co-Rotating and Counter-Roteting Twin-Screw Extruders using Mesh Partitioning Technique. In: ANTEC 2008 Plastics: Annual Technical Conference Proceedings, ANTEC, Milwaukee, 316-320.
18. Emin, M.A. and Schuchmann, H.P. (2013) Analysis of the Dispersive Mixing Efficiency in a Twin-Screw Extrusion Processing of Starch Based Matrix. Journal of Food Engineering, 115, 132-143.
19. Bravo, V.L., Hrymak, A.N. and Wright, J.D. (2000) Numerical Simulation of Pressure and Velocity Profiles in Kneading Elements of a Co-Rotating Twin Screw Extruder. Polymer Engineering and Science, 40, 525-541.
20. (2009) Ansys Polyflow 12.1 User's Guide.
21. Yacu, W.A. (1985) Modeling of a Twin Screw Co-Rotating Extruder.

Journal of Food Engineering, 8, 1-21.

22. Sakiyama, T., Han, S., Kincal, N.S. and Yano, T. (1993) Intrinsic Thermal Conductivity of Starch: A Model-Independent Determination. Journal of Food Science, 58, 413-415.
23. Noel, R.T. and Ring, S.G. (1992) A Study of the Heat Capacity of Starch/Water Mixtures. Carbohydrate Research, 227, 203-213. http://dx.doi.org/10.1016/0008-6215(92)85072-8.
24. Liu, J.W.H. (1992) The Multifrontal Method for Sparse Matrix Solution: Theory and Practice. Society for Industrial.
25. Rauwendaal, C. (2001) Polymer Extrusion. Carl Hanser Verlag, Munich.

Chapter 9

THE EFFECT OF SYSTEM BOUNDARIES ON THE MEAN FREE PATH FOR CONFINED GASES

Sooraj K. Prabha[1], Sreehari P. D.[1], Murali Gopal M.[1] And Sarith P. Sathian[1],

[1] Computational Nanotechnology Laboratory, National Institute of Technology Calicut, Kozhikode-673 601, Kerala, India

ABSTRACT

The mean free path of rarefied gases is accurately determined using Molecular Dynamics simulations. The simulations are carried out on isothermal argon gas (Lennard-Jones fluid) over a range of rarefaction levels under various confinements (unbounded gas, parallel reflective wall and explicit solid platinum wall bounded gas) in a nanoscale domain. The system is also analyzed independently in constitutive sub-systems to calculate the corresponding local mean free paths. Our studies which predominate in the transition regime substantiate the boundary limiting effect on mean free paths owing to the sharp diminution in molecular free paths near the planar boundaries. These studies provide insight to the transport phenomena of rarefied gases through nanochannels which have established their potential in microscale and nanoscale heat transfer applications.

INTRODUCTION

Recent advancements in micro/nanodevice technology connote the significance of nano-channels involving motion of fluids. These include mechanical systems like the micropumps, micro heat exchangers, microturbines, etc. In micro/nanoscale gas flows, the levels of rarefaction rely primarily on the non- dimensional Knudsen number (Kn), defined as the ratio of gas mean free path (λ) to the characteristic dimension (L). For a nano-sized channel, accurate determination of mean free path (MFP) over a range of rarefaction levels is one of the key parameters for the calculation of various molecular transport phenomena.

Accurate modeling of gas flows in nano sized structure is crucial for prediction of properties of these system. At the the molecular level computations MFP is an important parameter that governs the physical properties. [1] Generally, the flow properties are calculated by approximating the MFP using the theoretical models. [2,3] The theoretical models that dictate MFP calculate the bulk value and do not give any information regarding the variation with linear dimension, if any. [4,5] However there have been attempts to portray the variation of confined fluid with reflective wall boundary. [6,7] Furthermore, using the idea of local MFP there have also been attempts to predict the slip phenomenon. [8,9]

Dongari *et al.* have used the product of velocity after collision and time between adjacent collisions to calculate the distance between collisions for reflective wall confinements. Nevertheless, this approach fails when continuous potential fields are present as MFP is a function of position of molecules. When the gas is bounded with solid wall, the variation of number density of fluid atoms near the wall is significant. [10,11] In this study the attempt is to calculate the mean free path by tracking the trajectory of a molecule and detecting collisions.

In the present study, Molecular Dynamics (MD) method has been employed to calculate MFP of rarefied gas under different conditions. A single monatomic gas is at kept at a constant temperature and the MFP is obtained for different boundary conditions and rarefaction levels. The domain size and number density of atoms are selected so as to keep the rarefaction levels in the transition regime. In this study, the MFP is calculated by tracking the trajectory of each molecule and identifying collisions in the domain, which has periodic boundary conditions. The MFP thus obtained is compared with the MFP obtained by different mathematical models. Finally, The variation of MFP along the characteristic dimension for reflective and explicit boundary conditions is also compared at different rarefaction levels.

METHODOLOGY

A. Simulation model

A three dimensional system containing a monoatomic gas (argon) is the primary simulation domain. The interactions are modeled with Lennard-Jones (LJ) potential with a cut-off distance equal to three times the diameter of molecule. [12] Hard sphere diameter σ, is chosen to be 0.340 nm, where σ is the characteristic length parameter in LJ potential. The Ar-Ar interaction strength is taken to be 0.996 KJ/mol. [13] The fluid molecules are spatially distributed in a domain of size 1000 nm ^{3} with an initial Gaussian velocity distribution corresponding to a temperature of 300 K. For the unbounded gas, periodic boundary conditions (PBC) are employed in all three directions. The walls are modeled at either extremes of the domain in y direction (Figure. 1).

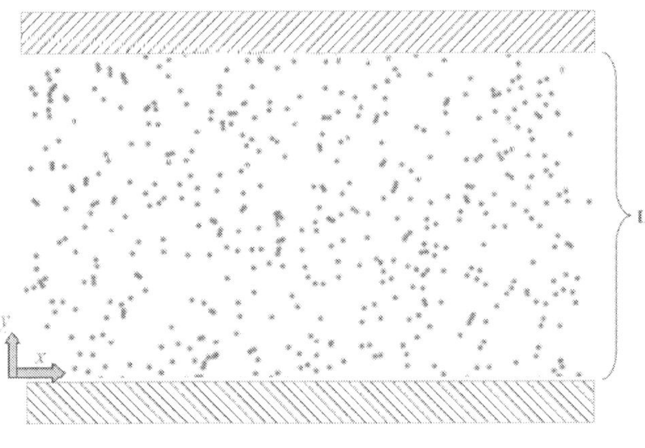

Figure 1. Skematic representation of the argon atoms and solid surface.

The characteristic dimension is also shown.

For the solid wall, Platinum atoms of thickness 1 nm are placed on either sides ensuring a parallel infinite wall boundary. The interaction strength for Pt - Pt interaction is taken to be 31.36 KJ/mol and for Pt - Ar interaction is taken to be 0.658 KJ/mol at a characteristic length of 0.294 nm. [13] The temperature of walls is maintained at 300 K using a Berendsen thermostat by controlling transitional degrees of freedom of wall atoms. The simulation domain is filled with argon gas with a number density ranging from 0.2 to 2.8 atoms/nm ^{3}. For the confined gas, this number density range corresponds to a Knudsen number range of 0.07 to 1.0, when the kinetic theory model is used to calculate the MFP. The total simulation run for data collection is for 50 ns. Velocity of the centre of mass of the system is periodically set to zero so as to nullify any net unbalanced force on the system. Parallel algorithm for classical MD with short range forces was used for simulations. [14]

B. Detecting collisions

Identification of intermolecular collisions is inevitable in calculating MFP. The time step size of 2 fs assures that all the collisions are captured. In the beginning, the system is allowed to relax to the equilibrium state. The position, velocity and timestep of the atoms are recorded at every timestep. A cubic collision domain of side length of about 3σ is chosen around every atom and all the atoms (neighbors) in the vicinity of the atom under consideration, within the collision domain are tested out for the occurrence of collisions. The intermolecular distance of σ or lesser registers a collision. For highly rarefied gases, since collisions are predominantly binary in nature, the odds of multiple simultaneous collisions between atoms are neglected.

C. Calculating molecular free path

The MFP of each molecule is calculated by averaging the distance they travel between collisions. In the case of simulation domains with PBCs, atoms escaping from the domains reappear at the opposite boundary. If a collision is detected immediately after reappearance, that collision is neglected for the calculation of MFP to avoid the error due to the imposed PBC. Thus, sufficient data is collected from the bulk fluid to get a stable calculated value of MFP.

RESULTS AND DISCUSSION

A. Unbounded gases

The ideal gas behavior is simulated by limiting the cut off distance to less than σ (3 Å) and thereby negating the influence of potential field during interaction. Accordingly well-established kinetic theory results can be compared with the simulation results and check the credibility of calculations. For this, argon gas at 300 K is simulated with periodic boundary conditions.

Velocity distribution

The velocity distribution obtained from the MD simulation of 2785 atoms at a temperature of 300 K is shown in Figure. 2 . When density and temperature are assigned to an ideal gas, then the velocity distribution will tend to the Maxwell's velocity distribution. [15]The velocity distribution obtained from simulation is conforming with the Maxwell's velocity distribution and it confirms the accuracy of the simulation model. As expected, the ideal behavior of argon gas is emulated in accordance with the kinetic theory and can be furthered to bounded gas simulations also.

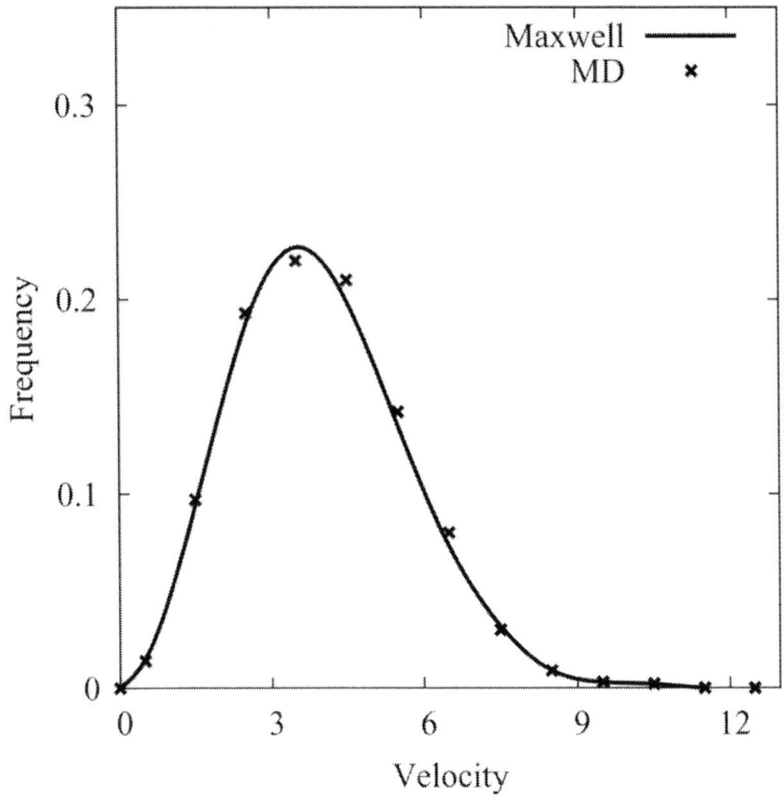

Figure 2. Comparison of Maxwell's velocity distribution and velocity distribution obtained from MD simulations at T=300 K, 2785 atoms

Mean free path of ideal gas

To further validate the MD model, MFP of the gas is calculated and compared to the theoretical values. From kinetic theory, the MFP of gases can be approximated to

$$\lambda_K = 1/\left(\sqrt{2}\pi d^2 n\right), \tag{1}$$

where d is the hard sphere diameter and n is the number density. The variation of MFP with the number density of atoms for the ideal gas ($\lambda\ Id$) is shown in Figure. 3 . The graph affirms that MFP is inversely proportional to the number density of the gas atoms. This shows that the simulation model could simulate ideal gas behavior and calculated MFP accurately matches with the MFP predicted with the kinetic theory approach ($\lambda\ K$). The MFP calculations are compared with another method of calculation, which uses velocity after collision and the time period between collisions ($\lambda\ VT$). [7] The values obtained from both the methods yield the same result and it shows that the trajectory tracing method can be effectively used even with PBC.

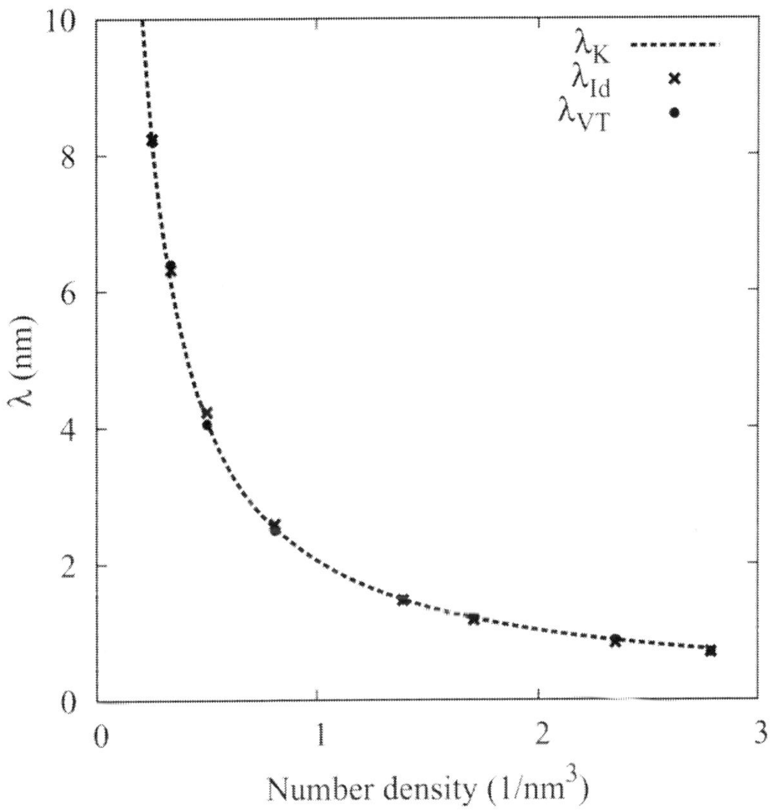

Figure 3. Variation of MFP with argon atom number density for unconfined ideal gas at 300 K. The curve represents the kinetic theory model ($\lambda\ K$) and the dots represent the results obtained from MD simulation ($\lambda\ VT$ and $\lambda\ Id$).

Furthermore, it can be observed that when the real gas behavior is introduced to the atoms, MFP computed from MD simulations (λRl) do not differ much from the ideal gas behavior for the selected rarefaction levels. The comparison of real gas behavior with ideal gas for bulk fluid is shown in Figure. 4 .

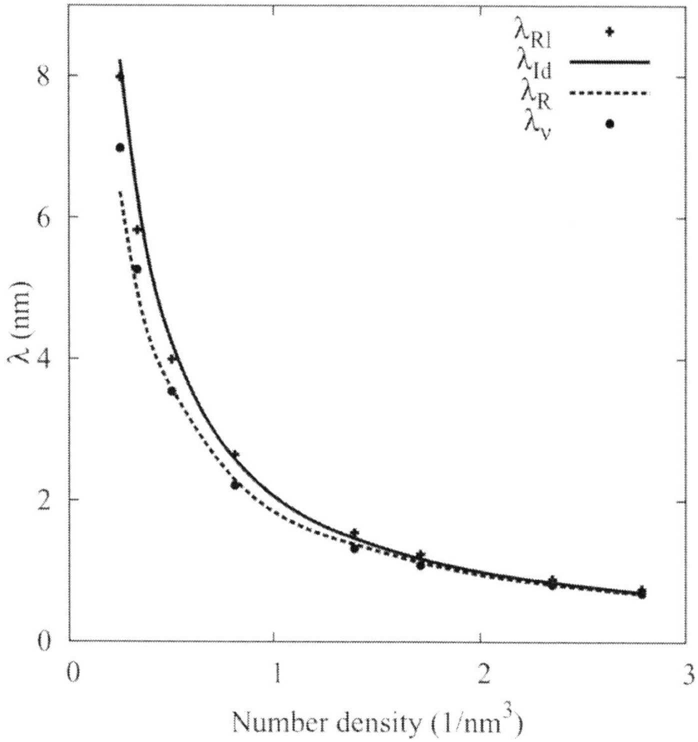

Figure 4. The comparison of MFP for ideal gas (λ *Id*), MFP for real gas (λ *Rl*), MFP for reflective wall boundary (λ *R*) and viscosity MFP (λ_v)

B. Ideal gas with reflective wall boundaries

The MFP of argon gas with reflective wall bounding surfaces is calculated to study the effect of wall boundary with MFP and its variation with rarefaction levels. It is observed that the average MFP shows a deviation from the theoretically calculated MFP and the deviation augments with increase in rarefaction (Figure. 4). The reflective wall changes the direction of normal component of velocity when the molecule crosses the boundary. This has two effects, the molecule crossing the boundary registers a collision and a slight increase in number density than the bulk portion as the molecules are redirected to the bulk fluid. Consequently, the presence of wall will increase the chances of fluid molecule undergoing a collision near the wall. This effect becomes more significant when the rarefaction levels of the fluid increases.

C. Viscosity MFP

Cercingani suggests that for particles with extended potentials, the mean free path can be expressed in terms of the viscosity by the following operational definition. [5,15] It is given as

$$\lambda_\nu = \frac{\nu}{\rho}\sqrt{\frac{\pi m}{2k_B T}}, \tag{2}$$

where k_B is the Boltzmann's constant and T is the temperature. The subscript v suggests that MFP is in terms of viscosity. Figureure4 shows the comparison of MFP for ideal gas (λ_{Id}), MFP for real gas (λ_{Rl}), MFP for reflective wall boundary (λ_R) and viscosity MFP (λ_v). The viscosity of argon gas was selected from experimentally obtained data at a constant temperature. [16] This suggests that the viscous effect imposed by the wall is noticeable as a net reduction in the effective mean free path of the gas at the bulk.

D. Variation of MFP along characteristic dimension

Figure 5 compares the variation of MFP along the characteristic length for a range of rarefaction levels for reflective and explicit wall models. The MFP in a bin is normalized with the average value of MFP obtained from MD. For better understanding of rarefaction level, the corresponding Knudsen number is also shown and Kn is calculated as, $Kn = 1/\left(\sqrt{2}\pi d^2 nL\right)$. Thus, for a moderately rarefied gas confined between smooth, flat and hard surfaces, viscosity MFP model can be effectively used to predict the MFP.

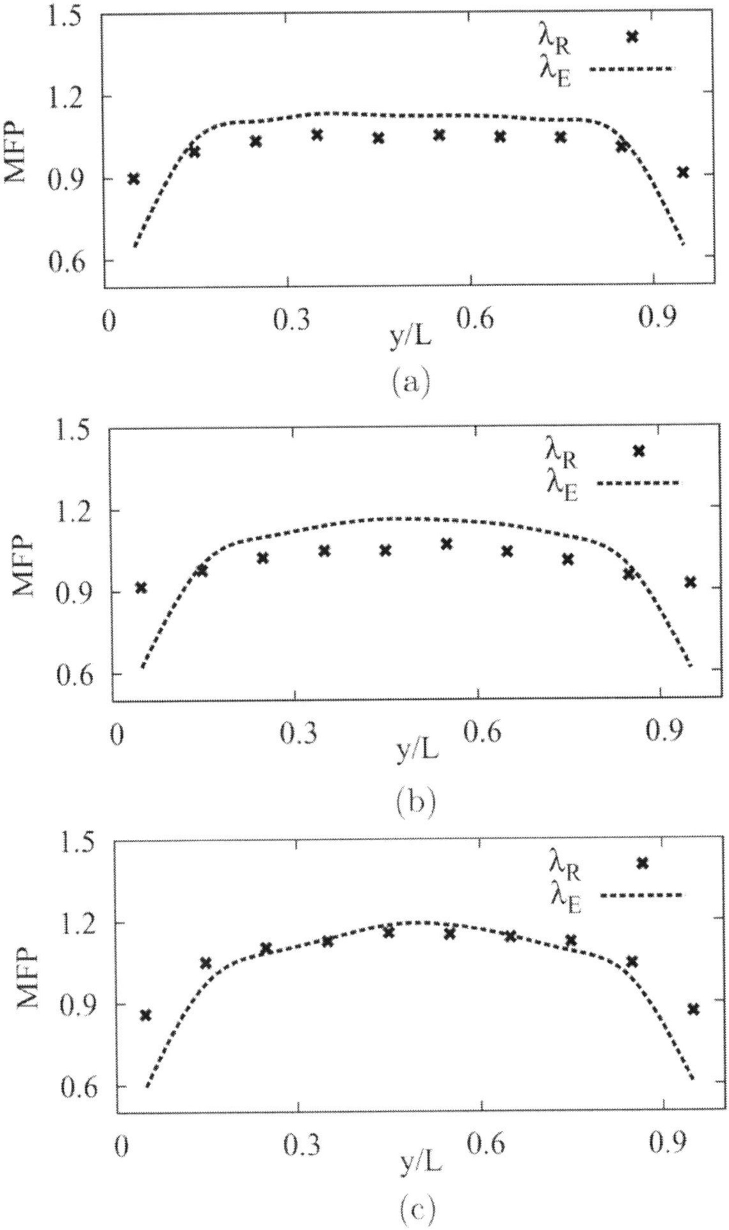

Figure. 5. Comparison of mean free paths obtained for confined gas under reflective wall (λ R) and solid platinum wall (λ E). (a) Kn = 0.15; (b) Kn = 0.4; (c) Kn = 0.8.

Reflective wall

It can be observed that the variation of normalized MFP shows similar variation with reflective wall boundary. For reflective wall, the MFP shows a slightly lower value than the bulk average near boundary. The fluid molecules near the wall are more probable to register a collision. However, it can be noticed that the variation of normalized MFP along characteristic dimension do not vary with the rarefaction levels.

Mean free path of real argon gas in solid platinum wall confinement

To study the effect of real wall environment the the fluid domain is provided with explicit wall boundary in the *y*-direction. The characteristic dimension is kept same as in the reflective wall confinement. In this case, simulations are carried out by considering solid-solid, solid-fluid and fluid-fluid interactions. The solid-solid interactions are limited to 2.5σ limit and all other interactions with 3σ limit. For explicit wall model, the collision with the wall is identified by defining a plane near the wall which is located at a molecular diameter distance away from the wall. When an atom crosses the plane towards the wall a collision is registered. When a gas atom interacting with the solid surface it is more probable to have multiple simultaneous collisions with the wall atoms. These simultaneous collisions, if any, are neglected; unless otherwise unrealistic values of MFP are obtained near the wall boundary.

For explicit wall model also shows a reduction in MFP near the wall ($\lambda\ E$). In this study, as the average MFP in the system domain is used to normalize the MFPs, there is a reduction in the MFP near the wall and subsequently the MFP in the middle portion is higher than the average value. For moderately dilute gases, corresponding to the transition regime, the reduction in MFP near the walls is significantly higher than that for the reflective wall boundary. This behavior can be attributed to the increase in the relative number of atoms near the surface as compared to that at the bulk. When number density decreases, fluid-fluid interaction decreases and the effect of surface potential is significant. This behavior becomes less significant as the gas atom number density increases while keeping all other parameters invariant.

It is interesting to note that the reduction in MFP near the wall, compared to the region away from the wall, is significant at lower number densities. As the rarefaction levels increases the Knudsen layer also increases and the variation of MFP extends deeper into the middle portion. Tang *et al.* [9] have also observed a similar variation of local MFPs between two plates. Apparently, it can be presumed that the variation in MFP is significant for the accurate determination of thermo-physical properties that are directly related to MFP.

CONCLUSION

Numerical Molecular Dynamics simulations are accomplished to accurately calculate the mean free path of argon gas in slip and transition regimes. The calculation of speed distribution and mean free path from the simulation of ideal gas behavior validates the accuracy of the MD model. The MFP is calculated by averaging the inter-collision distances of each molecule incorporating the effect of periodic boundary condition. The MFP calculated for real gases do not vary much from that of ideal gases at moderate rarefaction levels. Introduction of reflective wall boundaries results in the premature termination of atomic trajectories and hence MFP decreases near the boundary. The viscosity dependent MFP is in fair agreement with the results obtained from MD for reflective wall confinements. For explicit wall model, the variation of MFP is steeper near the boundary and it changes with number density.

REFERENCES

1. G. Karniadakis, A. Beskok, and N. R. Aluru, Microflows and Nanoflows (Springer, New York, 2005).
2. S. K. Prabha and S. P. Sathian, Phys. Rev. E **85**, 041201 (2012). http://dx.doi.org/10.1103/PhysRevE.85.041201
3. S. K. Prabha and S. P. Sathian, International Journal of Heat and Mass Transfer **58**, 217–223 (2013).http://dx.doi.org/10.1016/j.ijheatmasstransfer.2012.11.004
4. W.-M. Zhang, G. Meng, and X. Wei, Microfluidics and nanofluidics **13**, 845 (2012). http://dx.doi.org/10.1007/s10404-012-1012-9

5. D. L. Morris, L. Hannon, and A. L. Garcia, Physical review A **46**, 5279 (1992). http://dx.doi.org/10.1103/PhysRevA.46.5279
6. D. W. Stops, Journal of Physics D: Applied Physics **3**, 685 (1970). http://dx.doi.org/10.1088/0022-3727/3/5/307
7. N. Dongari, Y. Zhang, and J. M. Reese, Journal of Physics D: Applied Physics **44**, 125502 (2011).http://dx.doi.org/10.1088/0022-3727/44/12/125502
8. Z. Guo, B. Shi, and C. G. Zheng, EPL (Europhysics Letters) **80**, 24001 (2007). http://dx.doi.org/10.1209/0295-5075/80/24001
9. G. Tang, Y. Zhang, X. Gu, and D. Emerson, EPL (Europhysics Letters) **83**, 40008 (2008). http://dx.doi.org/10.1209/0295-5075/83/40008
10. S. K. Prabha and S. P. Sathian, Computers & Fluids **85** (2012).
11. S. K. Prabha and S. P. Sathian, Microfluidics and Nanofluidics **13**, 1–8 (2012).
12. M. P. Allen and D. J. Tildesley, Computer Simulation of Liquids (Oxford University Press, New York, 1994).
13. P. Spijker, A. J. Markvoort, S. V. Nedea, and P. A. J. Hilbers, Phys. Rev. E **81**, 011203 (2010).http://dx.doi.org/10.1103/PhysRevE.81.011203
14. S. J. Plimpton, J. Comp. Phys. **117**, 1 (1995). http://dx.doi.org/10.1006/jcph.1995.1039
15. S. Chapman and T. G. Cowling, The mathematical theory of non-uniform gases: an account of the kinetic theory of viscosity, thermal conduction and diffusion in gases (Cambridge university press, 1991).
16. J. Wilhelm and E. Vogel, International Journal of Thermophysics **21**, 301 (2000).http://dx.doi.org/10.1023/A:1006667125801

Chapter 10

COMPUTATIONAL FLUID DYNAMICS AND ITS IMPACT ON FLOW MEASUREMENTS

Claus Kiefer*, Frauke Kellner-Weldon

Support Center for Advanced Neuroimaging, Institute of Diagnostic and Interventional Neuroradiology, University of Bern (Inselspital), Bern, Switzerland

ABSTRACT

Rationale and Objectives: Computational fluid dynamic (CFD) simulations are discussed with respect to their potential for quality assurance of flow quantification using commercial software for the evaluation of magnetic resonance phase contrast angiography (PCA) data. Materials and Methods: Magnetic resonance phase contrast angiography data was evaluated with the Nova software. CFD simulations were performed on that part of the vessel system where the flow behavior was unexpected or non-reliable. The CFD simulations were performed with in-house written software. Results: The numerical CFD calculations demonstrated that under reasonable boundary conditions, defined by the PCA velocity values, the flow behavior within the critical parts of the vessel system can be correctly reproduced. Conclusion: CFD simulations are an important ex tension to commercial flow quantification tools with regard to quality assurance.

INTRODUCTION

Magnetic resonance phase contrast angiography mean- while is an established technique to measure the velocity and the direction of flow [1]. The major drawbacks of this approach are often related to misaligned encoding gradients (venc), a too low SNR or high susceptibility gradients nearby (small) vessels. In each case the relative phase are not unique and the absolute velocity values are wrong. At this point the correct answer to the flow in the critical regions can only be derived on the base of physical models and computational flow dynamics (CFD). Almost all CFD-problems are associated with the solution of the Navier-Stokes equations [2,3], named after Claude-Louis Navier and George Gabriel Stokes, which describe the motion of fluid substances. The calculations were performed on incompressible fluids within human blood vessels. The idea is to selectively model the relevant vessel system and to include the known velocity values from the PCA measurements within the big vessels as boundary conditions. These velocity conditions are sufficient in most cases because the velocities tune in to the given pressure conditions. The intrinsic symmetry of the vessels furthermore can be used to reduce the dimensionality of the problem from three to two, which is necessary to use the software on commercial computers with limited working memory and CPU power—only the relative orientation of the vessels and their diameters (e.g. due to stenoses) have to be considered. The spectrum of potential applications span the simulation of pressure and force distributions near the neck of aneurysms, the velocity and direction of flow within complex vessel systems with or without stenoses and pressure and vorticity distributions near atherosclerotic pla-ques. In this technical note we will concentrate on a few representative examples in order to demonstrate the importance of quality assurance in addition to commercial flow evaluation software.

METHODS

Flow Measurement

Magnetic resonance phase contrast angiography measurements of flow were performed on a 3 Tesla scanner. The Flash sequence parameters were as follows: Base resolution 256, Phase resolution 60%, Flip angle 25 deg, TR 113.20 ms, TE 4.66 ms, Voxel size: 0.9 × 0.5 × 4.0 mm, TA: 1:06 min, 1st Signal/Mode Pulse/Retro, calculated phases 12, segments 5, slices 1. Each slice was planned on the base of a time-of-flight (TOF) measurement using the Nova software (http://www.vassolinc.com), which guarantees the exact orthogonal positioning relative to the course of the vessel. The flow measurements were performed for a number of selected vessels that define a part of the simulated vessel system. Figure 1 shows a typical Nova rendered 3D-presentation of the vessel system of a TOF measurement. The velocities are measured at the positions given by the small slice cartoons perpendicularly oriented to the vessel. The most important aspects which are essential for the choice of these locations are the distance to bifurcations and stenoses because often turbulences may occur in their immediate vicinity. Nova currently provides no option for the separate presentation of systolic and diastolic phases but merely shows the mean volumetric flow rate values on the diagrams (Figure 1(b)). The mean velocity values, necessary for the simulations, are not shown on the Nova diagram, but are stored in a text file.

Preparation of CFD

2D-projections from individual TOF maximum intensity projection maps are co-registered to a template (see e.g. http://commons/Wikimedia.org/wiki/Template:Other_versions/Circle_of_Willis, binary mask in Figure 2(a)) while preserving the individual properties (e.g. stenoses, mal- formations, relative orientation). The final mask was used to define the coordinates and nodes essential for the mesh2d Matlab program, which calculates the nodes and elements (Figure 2(b)) necessary for the numerical simulations.

Numerical Algorithms

The numerical code for the flow simulations is an in-house development, written in Matlab. The program simulates the motion of incompressible fluids via numerical solutions of the 2D, unsteady Navier-Stokes equations. The CFD-program is a vertex centered Finite Volume (FV) code that uses the median dual mesh to form the control volumes about each vertex by connecting the centroids of each cell to their edge midpoints. The time and spatial linearizations are dealt with separately. Fully 2nd order piecewise linear reconstructions are used to develop sparse operators, which approximate the continuous differential terms. These methods can be found in detail in [4, 5] described by terms like Green's theorem and its application to face averaged gradients (Laplacian, un-weighted least squares) or upwind piece-wise linear extrapolation with the HQUICK slope limiter (non-linear part). The time integration of the equations is based on the well-known fractional step method [5,6]. The non-staggered pressure/velocity arrangement was shown to support the common mesh scale pressure oscillation and hence the standard 2nd order incremental pressure correction method [7,8] was modified to incorporate Rhie-Chow stabilization [9]. This reduced the order of the time integration to 1st/2nd. Unlike many other Navier-Stokes solvers the CFD-program makes use of a complete LU factorizetion for the Poisson pressure equation. Instead of using standard iterative approaches like preconditioned bi-conjugate gradients stabilized method (bicgstab (Matlab)) we decided on including the fill reducing UMFPACK algorithm [10], an approach which was shown to be far more efficient in this case. The momentum equations are solved in a semi-implicit manner. The non-linear terms are evaluated explicitly via a 2nd order TVD Runge-Kutta method [11]. In order to solve the linear systems that arise due to the implicit treatment of the viscous terms, the bicgstab method was favored. Minimization of errors associated with the numerical solution was managed ac- cording to the concepts of Roache et al. [12].

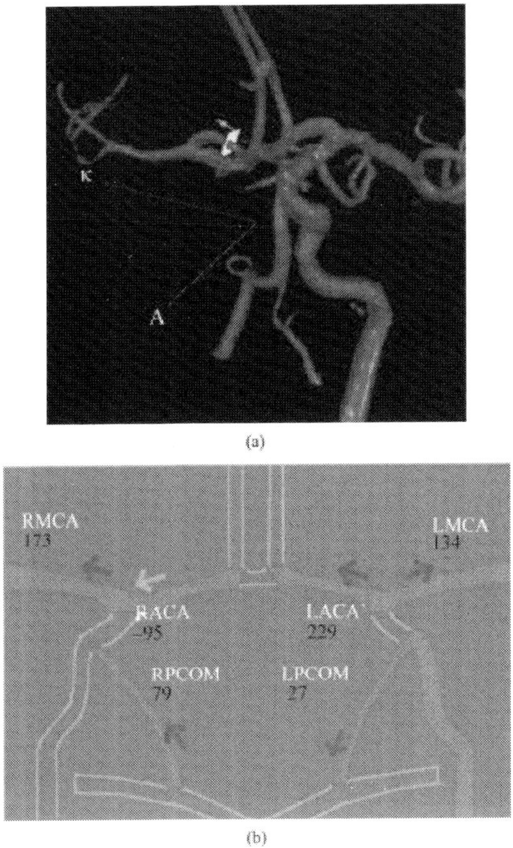

Figure 1. A typical 3D-plot (a) of a NOVA flow planning system, illustrating the rendered vessel system according to the TOF measurement. The velocities are measured at the positions given as small slice cartoons perpendicularly oriented to the vessel (here the right ACA). In (b) one possible schematic Nova diagram of a part of the circle-of-willis system is shown, which contains information about the vessel names, the direction of flow (green arrow) and volumetric flow rate values (ml/min).

CPU Parameters

The CFD simulations were performed on a personal computer with Intel® Xeon processor, 2.4 GHz, 12 GB RAM, CPUs, NVIDIA Quadro FX 580, Video RAM 512 MB, GPU frequency 450 MHz, Ubuntu 10.04.

Program Features

The mesh2d program of Mat lab was used for the 2D mesh (triangles) generation. This type of mesh allows the automatic treatment of complex geometries. The mesh2d algorithm optimizes the mesh quality which is important to guarantee correct results. The CFD program solves the Navies-Stokes equations for the velocity and pressure fields of incompressible fluids. The time step size and the stability of the integration are adjustable. The velocity values measured by PCA were used as the boundary conditions (at the nodes). The program includes a module that calculates the [x,y] forces applied to a boundary due to pressure and skin friction effects and can be used in this example to calculate the lift and drag forces on an aneurysm neck.

Patient

High grade stenosis of the left ICA and occlusion of the right ICA.

RESULTS

The total time needed for the preparation and simulation is in the range of 3 min which is actually feasible for emergency cases. The computer, however, should at least have 4 GB working memory and processors of the new generation (Pentium IV, Xeon) with at least 2.4 GHz. Figure 2(a) shows a typical bitmap of the circle-of- willis which is then transformed to a 2D-mesh with nodes and elements (Figure 2(b)) using the mesh2d Mat-lab program. As a typical example for the application of the CFD-simulations, the data of a patient before and after stenting of the left ICA was used. The velocity values at the boundaries were given by the PCA measurements. The simulations of the corresponding vessel sys-tem before (Figures 3(a)-(c)) and after stenting (Figures 4(a)-(c)) confirm the change in direction within the LPCOM vessel (Figure 4(b)) with respect to the situation before stenting (Figure 3(b)): it is opposite to the flow within the RPCOM (Figures 3(c) and 4(c)). The associated Nova diagrams are shown in Figure 3(d) (be- fore stent) and Figure 4(d) (after stent). The measured mean velocity values are shown in Table 1.

In order to demonstrate the potential of the CFD simulation

technique, an example for such an (artificial) aneurysm is given in Figure 5, for which the velocity (Figure 5(a)) and pressure distributions (Figure 5(b))

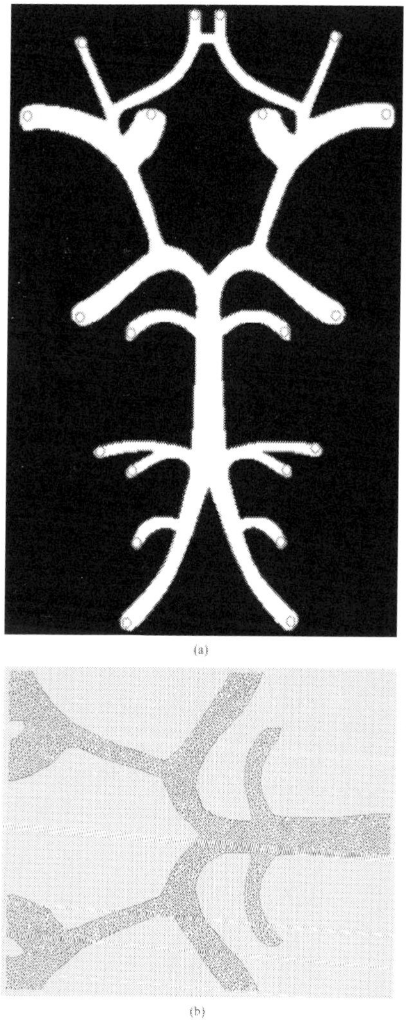

Figure 2 The bitmap of the circle-of-Willis vessel system (a) and the mesh nodes and elements for a part of it (b), calculated by the mesh2d Mat lab program. The mesh serves as the base for the CFD simulations. Potential boundaries which may have non-zero in- or outflow are marked with red cycles.

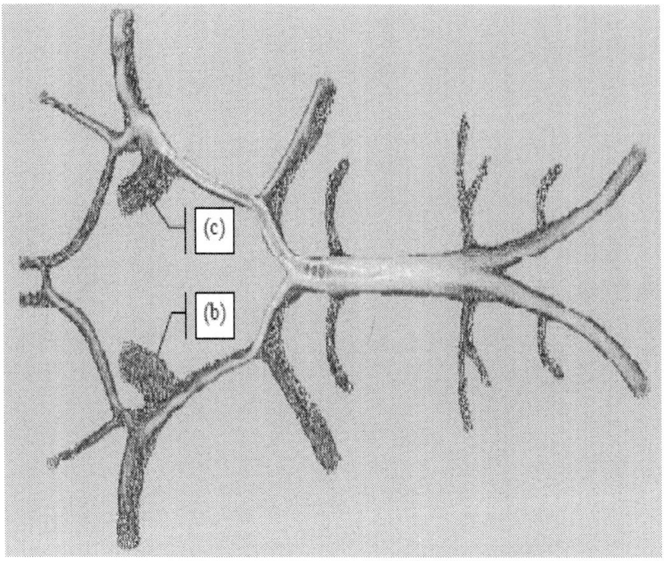

(a)

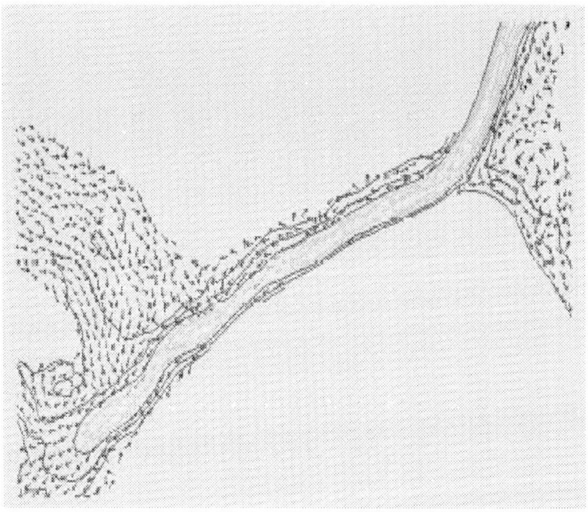

(b)

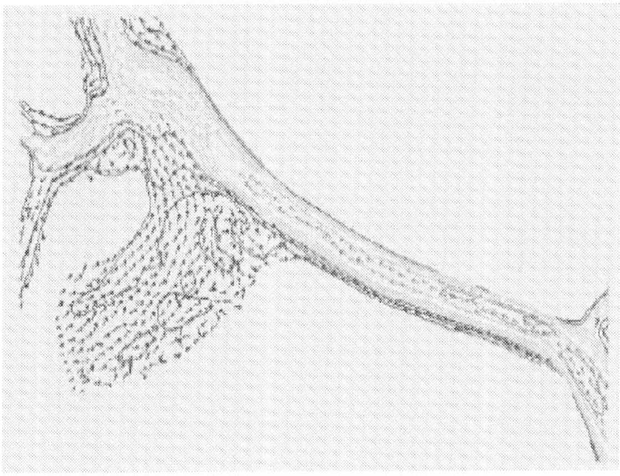

(c)

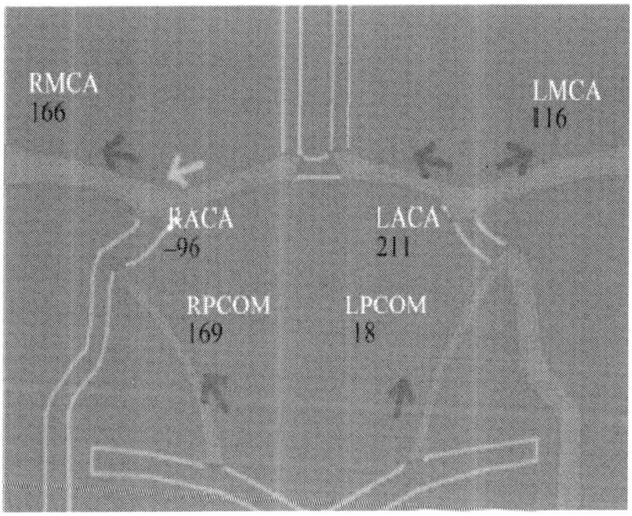

(d)

Figure 3 Simulated CFD data of the circle-of-Willis vessel system, shown in Figure 2, if the boundary conditions are given by the velocity values of a real PCA measurement be- fore stenting: note the equal flow directions in the LPCOM (b) and RPCOM (c) which coincides with the results of the Nova data (d).

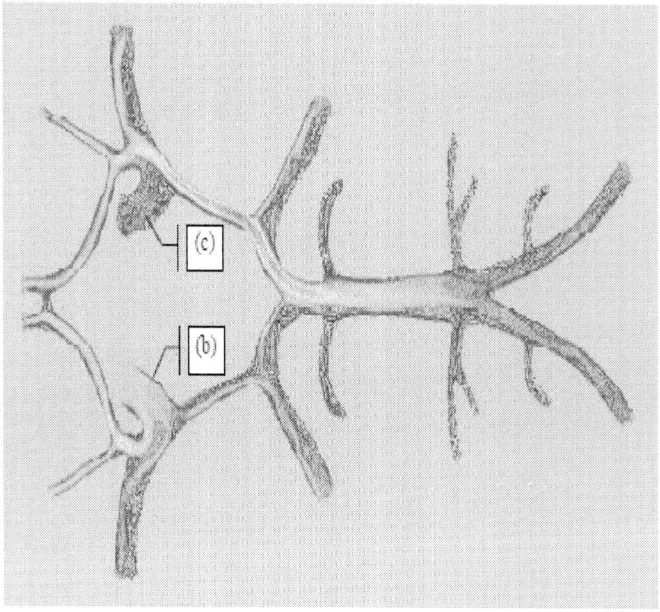

(a)

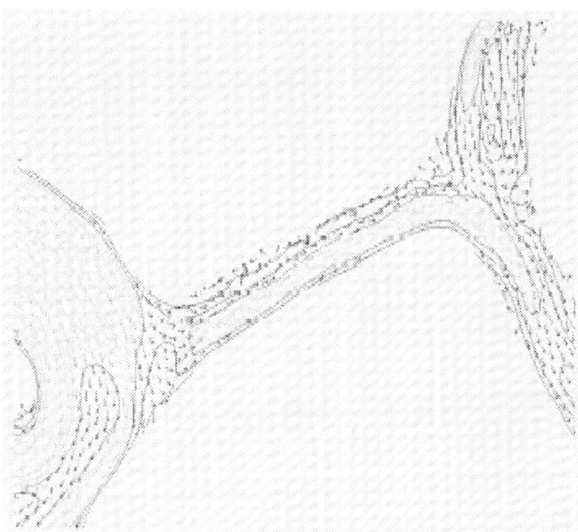

(b)

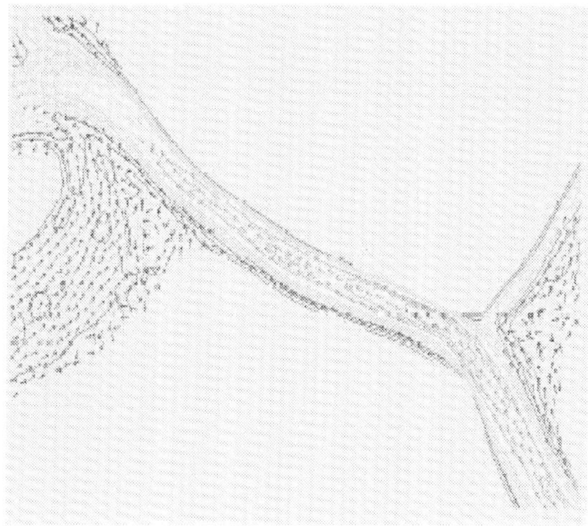

(c)

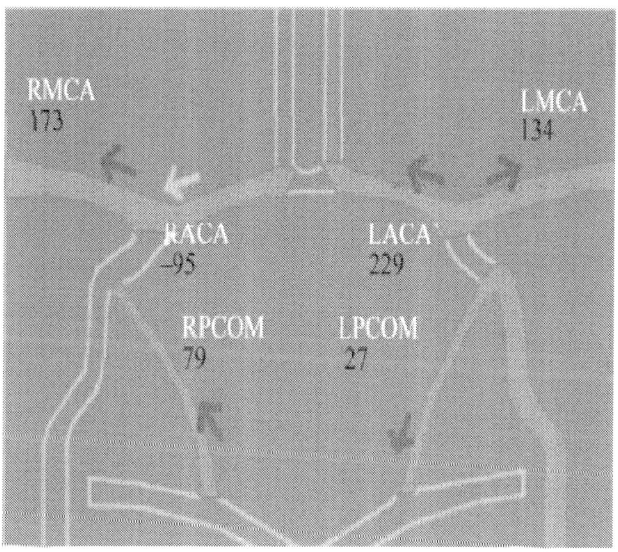

(d)

Figure 4 Simulated CFD data of the circle-of-Willis vessel system, shown in Figure 2, if the boundary conditions are given by the velocity values of a real PCA measurement after stenting: note the opposite flow directions in the LPCOM (b) and RPCOM (c) and the change in direction within the LPCOM vessel (Figure 4(b)) with respect to the situation before stenting (Figure 3(b)) which coincides with the results of the Nova data (d).

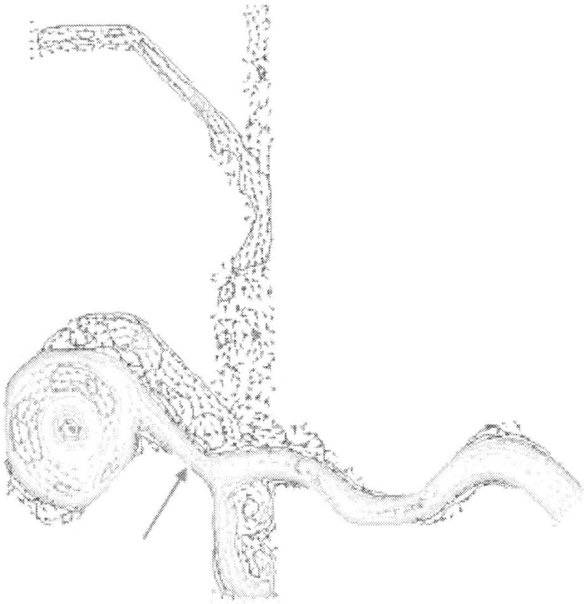

(a)

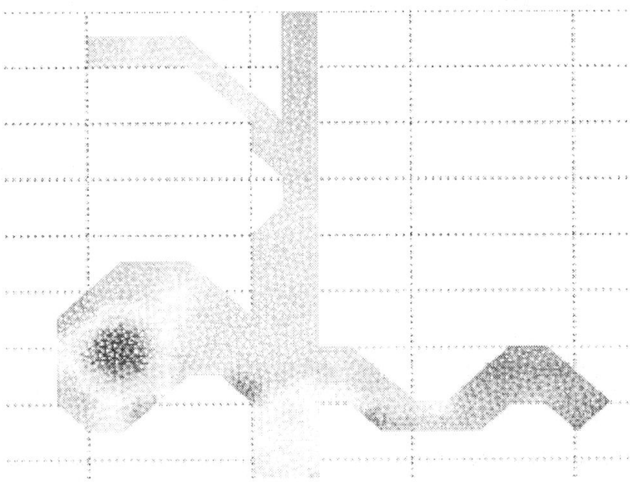

(b)

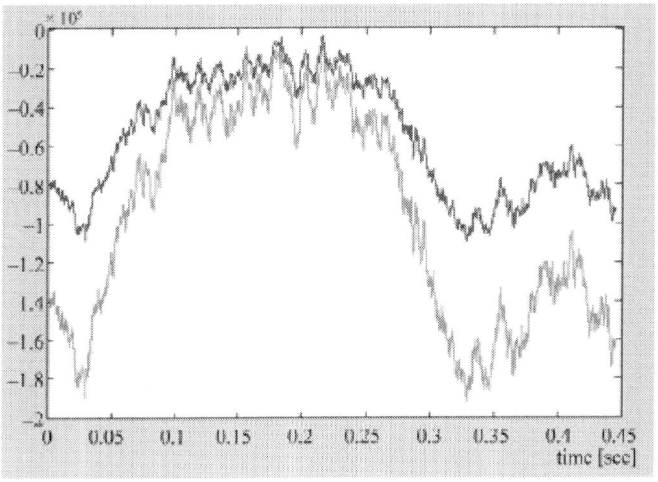

(c)

Figure 5 CFD simulations of an artificial aneurysm. (a) The pressure distribution (b) and the forces (in dependence of time [sec]) (c) at the marked position (red arrow in a)) demonstrate, that under special conditions the local load can achieve critical values, followed by a disruptor of the aneurysm neck.

Table 1 Mean velocities within selected vessels in cm/sec.

Vessel	LACA	RACA	LPCOM	RPCOM	LMCA	RMCA
Before	42.61	13.35	7.93	42.77	26.32	30.47
After Stent	55.70	29.29	10.71	27.62	30.88	29.62

are shown if physiologically reasonable boundary conditions are presumed. At the border of the area, indicated by the red arrow, the forces (**Figure 5(c)**) in x – (blue) and y – direction (red) are shown in dependence of time. The absolute values are not relevant in this case but merely the relative pressure and force enhancements at the neck of the aneurysm which finally may lead to a disruptor.

DISCUSSION

CFD simulations were used to verify unexpected flow patterns that occur after stenting in the LPCOM vessel and to demonstrate the critical pressure and force distributions at the neck of an artificial aneurysm. The data shows that CFD simulations are feasible on commercially available PCs if the simulations are restricted to the essential vessel systems and if intrinsic symmetries are used to reduce the dimensionality of the problem. CFD data may also provide clinically useful information regarding post-stenotic vessel wall alteration and progression of atherosclerotic disease. More generally, wall shear stress is believed to play an important role in the evolution of atherosclerosis and other arterial diseases, such as aneurysms.

At the current stage only vortices but no turbulence model is implemented. Further CFD program options are different boundary conditions such as to extrapolate the boundary values of the pressure from the interior nodes or just to fix the value of the pressure at the outflow to zero.

The step from the TOF-MIP to a two-dimensional bit- map is currently not fully automated but necessitates manual intervention by an experienced radiologist.

In future developments, a separate evaluation of different heart phases will be possible as well. This deficit, however, does not lower the impact of the results, especially because the flow directions before and after stenting do not depend on the heart phase.

In summarize, the intention of this work was to demonstrate, that the simulation of model based fluid dynamics provides important clinical information which is currently not available in most commercial software packages for flow quantification.

REFERENCES

1. C. L. Dumoulin, S. P. Souza, M. F. Walker and W. Wagle, "Three-Dimensional Phase Contrast Angiography," Mag- netic Resonance in Medicine, Vol. 9, No. 1, 1989, pp. 139-149. doi:10.1002/mrm.1910090117
2. G. K. Batchelor, "An introduction to Fluid Dynamics," Cambridge University Press, Cambridge, 2000.

3. E. Erturk, T. C. Corke and C. Gökçöl, "Numerical Solu- tions of 2-D Steady Incompressible Driven Cavity Flow at High Reynolds Numbers," International Journal for Nu- merical Methods in Fluids, Vol. 48, No. 7, 2005, pp. 747- 774. doi:10.1002/fld.953
4. D. Kim and H. Choi, "A Second-Order Time Accurate Fi- nite Volume Method for Unsteady Incompressible flow on Hybrid Unstructured Grids," Journal of Computatio- nal Physics, Vol. 162, No. 2, 2000, pp. 411-428. doi:10.1006/jcph.2000.6546
5. J. Kim and P. Moin, "Application of a Fractional Step Method to the Incompressible Navier-Stokes Equations," Journal of Computational Physics, Vol. 59, No. 2, 1985, pp. 308-323. doi:10.1016/0021-9991(85)90148-2
6. A. Chorin, "Numerical Solution of the Navier-Stokes Equa- tions," Mathematics of Computation, Vol. 22, No. 104, 1968, pp. 745-762. doi:10.1090/S0025-5718-1968-0242392-2
7. S. Armfield and R. Street, "Modified Fractional Step Me- thods for the Navier-Stokes Equations," ANZIAM Journal, Vol. 45, 2004, pp. 364-377.
8. S. Armfield and R. Street, "The Pressure Accuracy of Fractional Step Methods for the Navier-Stoke Equations on Staggered Grids," ANZIAM Journal, Vol. 44, 2003, pp. 20-39.
9. C. Rhie and R. Chow, "Numerical Study of the Turbulent Flow Past and Airfoil with Trailing Edge Separation," The American Institute of Aeronautics and Astronautics, Vol. 21, No. 11, 1983, pp. 1525-1532. doi:10.2514/3.8284 [10] http://www.cise.ufl.edu/~davis/
10. C. W. Shu, "Total Variation Diminishing Time Discretisa- tions," SIAM Journal on Society and Scientific Comput- ing, Vol. 9, No. 6, 1988, pp. 1073-1084. doi:10.1137/0909073
11. P. J. Roache, "Quantification of Uncertainty in Computa- tional Fluid Dynamics," Annuls Reviews of Fluid Me- chanics, Vol. 29, 1997, pp. 123-160. doi:10.1146/annurev.fluid.29.1.123

Citations

CHAPTER 1

W. Limtrakarn, P. Boonmongkol, A. Chompupoung, K. Rungprateepthaworn, J. Kruenate, and P. Dechaumphai, "Computational Fluid Dynamics Modeling to Improve Natural Flow Rate and Sweet Pepper Productivity in Greenhouse," Advances in Mechanical Engineering, vol. 2012, Article ID 158563, 7 pages, 2012. doi:10.1155/2012/158563

CHAPTER 2

Vadim Seleznev (2010). Computational Fluid Dynamics Methods for Gas Pipeline System Control, Computational Fluid Dynamics, Hyoung Woo Oh (Ed.), ISBN: 978-953-7619-59-6, InTech, DOI: 10.5772/7110.

CHAPTER 3

Rabindra Nath Mondal, Samir Chandra Ray, Shinichiro Yanase, Combined Effects of Centrifugal and Coriolis Instability of the Flow through a Rotating Curved Duct with Rectangular Cross Section, DOI: 10.4236/ojfd.2014.41001

CHAPTER 4

Li Li, "An Initial Investigation of Adjoint-Based Unstructured Grid Adaptation for Vortical Flow Simulations," Modelling and Simulation in Engineering, vol. 2011, Article ID 217326, 9 pages, 2011. doi:10.1155/2011/217326

CHAPTER 5

Carlos Gavilán Moreno (2010). Turbulence, Vibrations, Noise and Fluid Instabilities. Practical Approach., Computational Fluid Dynamics, Hyoung Woo Oh (Ed.), ISBN: 978-953-7619-59-6, InTech, DOI: 10.5772/7100.

CHAPTER 6

Ki-Pyo You and Young-Moon Kim, "Wind Velocity Decreasing Effects of Windbreak Fence for Snowfall Measurement," Advances in Meteorology, vol. 2014, Article ID 791256, 20 pages, 2014. doi:10.1155/2014/791256

CHAPTER 7

Mohammad Bakhtiari, Hassan Ghassemi, Numerical Analysis of Electromagnetic Control of the Boundary Layer Flow on a Ship Hull, dx.doi.org/10.4236/ojfd.2014.41006

CHAPTER 8

A. Cubeddu, C. Rauh and A. Delgado, "3D Thermo-Fluid Dynamic Simulations of High-Speed-Extruded Starch Based Products," Open Journal of Fluid Dynamics, Vol. 4 No. 1, 2014, pp. 103-114. doi: 10.4236/ojfd.2014.41008.

CHAPTER 9

Sooraj K. Prabha1, Sreehari P. D.1, Murali Gopal M.1 and Sarith P. Sathian1, The effect of system boundaries on the mean free path for confined gases, dx.doi.org/10.1063/1.4824634.

CHAPTER 10

Claus Kiefer, Frauke Kellner-Weldon Support Center for Advanced Neuroimaging, Institute of Diagnostic and Interventional Neuroradiology, University of Bern (Inselspital), Bern, Switzerland, Computational Fluid Dynamics and Its Impact on Flow Measurements Using Phase-Contrast MR-Angiography, DOI: 10.4236/ojmi.2012.21004

INDEX

P

R

S

T

U

V

W